Middle East Today

The Iranian Revolution of 1979, the Iran-Iraq War, the Gulf War, and the US invasion and occupation of Iraq have dramatically altered the geopolitical landscape of the contemporary Middle East. The Arab Spring uprisings have complicated this picture. This series puts forward a critical body of first-rate scholarship that reflects the current political and social realities of the region, focusing on original research about contentious politics and social movements; political institutions; the role played by non-governmental organizations such as Hamas, Hezbollah, and the Muslim Brotherhood; and the Israeli-Palestine conflict. Other themes of interest include Iran and Turkey as emerging pre-eminent powers in the region, the former an 'Islamic Republic' and the latter an emerging democracy currently governed by a party with Islamic roots; the Gulf monarchies, their petrol economies and regional ambitions; potential problems of nuclear proliferation in the region; and the challenges confronting the United States, Europe, and the United Nations in the greater Middle East. The focus of the series is on general topics such as social turmoil, war and revolution, international relations, occupation, radicalism, democracy, human rights, and Islam as a political force in the context of the modern Middle East.

Rawan Asali Nuseibeh

Urban Youth Unemployment, Marginalization and Politics in MENA

palgrave
macmillan

Rawan Asali Nuseibeh
Institute of Educational Research
Addis Ababa University
Addis Ababa, Ethiopia

ISSN 2945-7017 ISSN 2945-7025 (electronic)
Middle East Today
ISBN 978-3-031-15300-6 ISBN 978-3-031-15301-3 (eBook)
https://doi.org/10.1007/978-3-031-15301-3

Cover credits: Claudia Wiens/Alamy Stock Photo

This Palgrave Macmillan imprint is published by the registered company Springer Nature Switzerland AG
The registered company address is: Gewerbestrasse 11, 6330 Cham, Switzerland

Contents

CHAPTER 1

Introduction

The accelerated urbanization happening worldwide is accompanied with urban social segregation and inequality. Inhabitants of cities do not have equal opportunities, whether in terms of access to services, housing, safety, transportation or in terms of having a say in how their space is shaped. The Arab region is one of the most urbanized regions in the world, with around 60% of its inhabitants living in cities (UN habitat, 2013). However, colonization, internal migration and exponential growth rate, have resulted in unplanned and poorly structured growth in Arab cities, resulting in huge gaps between formally designed neighborhoods and informal settlements known as *Ash-wa'iyyat* or slums. However, in the case of East Jerusalem, the problem is not informal or haphazard planning of the city, it is actually the opposite, as explained by several urban geographers; the planning of the city is engineered towards ensuring a hegemonic Israeli Jewish identity, while limiting and erasing the Palestinian identity (See AlKhalili, 2019; Shtern, 2018; Shlomo, 2016; Yiftachel & Yacobi, 2006). This book will look into the lives of Arab youth in four cities Cairo, Tunis, Amman and Jerusalem.

What is common in our four contexts is the exclusion experienced by youth residing in marginalized neighborhoods. Arab youth residing in marginalized neighborhoods are harmed by the lack of opportunities available to them. They dwell in neighborhoods that lack basic social

R. A. Nuseibeh, *Urban Youth Unemployment, Marginalization and Politics in MENA*, Middle East Today,
https://doi.org/10.1007/978-3-031-15301-3_1

services, proper housing, safe environments and transportation, which adversely affect their life trajectories and well-being. Arab youth, in particular, are mostly excluded from decision-making apparatuses, which decide how cities are shaped. They are treated as either a problem and referred to as the "youth bulge" or a security threat "radicalized youth", or as the "hope of the nation" (Calder et. al.2017:11). There are some initiatives that include youth in decision-making processes, but they remain either theoretical discussions or very limited in scope. The dominant frameworks used to study Arab youth are mostly economic ones, that focus on unemployment and the human capital aspects.The narrative around Arab youth has been narrowed down to economic aspects of unemployment and precarity, focusing solely on human capital in the region and how to improve the levels of human capital, in order to reduce unemployment rates. However, in Arab countries, it is hard to make the case that improving human capital is a solution to youth exclusion, since unemployment is very high among young educated Arabs, also, Arab women are more likely to have a university degree and are still lagging behind in labour market participation.

Therefore, overall macro-approaches and utilitarian principles, such as traditional human development approaches, cannot provide the necessary framework to understand youth exclusion in the region (Lopez-Fogues & Melis Cin, 2017). This book will use a combination of Lefebvre's (1968) "right to the city" and Sen's (1999,1992,1985) "capability approach" frameworks, to study the lives of Arab youth in the four cities. The use of those two frameworks to study youth exclusion in cities is not new. Deneulin (2014) has used these two frameworks in the context of Latin American cities. She argues that the "capability approach" fills the gaps in the "right to the city" approach. By using the two frameworks of the "capability approach" and the "right to the city" approach, we can see what a "just city" for all looks like (Fainstien, 2010) and how the current reality of youth in Arab cities does not reflect a just city for all, instead, urban spaces are becoming increasingly commercialized and the power to shape the cities is in the hands of the few.

For this book, I have conducted several interviews with civil society organizations that work on youth programs, as well as interviews with youth residing in the four cities, ages between 19 and 34. The sample was diverse, it included people at various stages of their careers from various professions, students and unemployed. It also included an almost equal number of males and females. The majority was Muslim with a few Christian participants.

Chapter 1 starts by an introduction about the book, its major theoretical framework and the division of its chapters and topics. Chapter 2 starts by exploring the combined theories of the "right to the city" and the "capabilities approach." It then looks at how our four cities are shaped, how they became to be what they are today, and how life in those cities is divided and experienced by different groups. It explores how youth are affected by the various policies in the cities, and how much of a say they have in shaping their cities. It examines how those cities are governed, the level of people's engagement in local elections and how fair and democratic those local elections are. Chapter 3 looks at the political and economic climates that shape youth opportunities in the city, it also looks at employment prospects and opportunities in the labour market, marginalization within the urban space and how the economic and political climate in their countries are leading to their dependence on and subordination to elders in their family units. Chapter 4 examines educational services offered. It starts by setting out the utopian vision of an education system in the city that aims at building people's capabilities, enhancing individual empowerment, achieving sustainable economic development, while increasing community engagement and active citizenship. It calls for a paradigm shift in the view of educational policy from focusing solely on economic returns of education, to look at education through the lens of the "capability approach"; in order to see to what extent education builds people's capabilities and builds an empowered sense of agency. It also calls for an education system that serves all city residents regardless of social class, ethnicity and gender. To create this kind of education system, residents of the city need to be able to participate in decisions affecting their lives in the urban space, including the distribution of resources towards educational institutions, and how these institutions function and how inclusive these institutions are for all the communities residing in that space. The chapter then brings in the concept of the "right to the city", which seeks to engage members of the community democratically in the shaping of their educational institutions, and the changing of culture towards principals of human flourishing and enhancement of capabilities and social justice. The chapter then explores the actual state of the education sector in our four contexts, and reveals the rife inequalities and deficiencies that can suppress capabilities, rather than build them. The chapter also looks into the governance of education systems in the four contexts, looking in particular at issues of democratic governance and transparency. Chapter 5 looks at higher education in our four contexts, and the two-way relationship between the university and the city; while institutions of higher education contribute to cities, on the economic, political and intellectual and cultural levels, the cities' infrastructure, political and economic stability also increase demand and attract cadres to their universities. The importance of institutions of higher education on the city and the country level means that these institutions can have the power to either reproduce hegemonic powers and entrench inequalities, or be the power to change and give voice to the marginalized. Therefore, their governance is as important as the governance of cities. The political subordination of institutions of higher education to authoritarian governments, lack of transparency and lack of independence, affect how these institutions function and what their aims are.

The final chapter 6 moves from the analysis of the structures that shape the lives of youth to explore youth agency and how they shape their cities, and resist the current realities they are living in. Although Arab youth's access to formal power is marginal, they are not passive actors in their communities and they have shown their power to revolt against structural injustices, as was reflected in the revolts that shook the streets of Arab cities. Various grassroots organizations, civil-based institutions and groups have managed to organize the fight for residents' rights; some groups advocate and act alone, while others have managed to build bridges and construct alliances with other groups within the city to envision alternative possibilities, and fight for a more just city that serves all (Miller & Nicholls, 2013). The chapter also looks at the role of digital activism and the wide reach of media, and how that has enabled civil groups to support each other across borders.

The book concludes by proposing a re-imagining of the city through the lenses of the "right to the city" and the "capabilities approach" frameworks; imagining cities as not only fulfilling the civil, political and social rights of all residents, but also as empowering city residents to participate in the shaping of their cities. It calls for re-imagining the governance of the cities, the role of political institutions, and the role of educational institutions at various levels from pre-primary to higher education and re-imagining the labour markets and the political environments in which these markets function.

Bibliography

Alkhalili, N. (2019). 'A forest of urbanization': Camp Metropolis in the edge areas. *Settler Colonial Studies, 9*(2), 207–226.

Calder, M., MacDonald, R., Mikhael, D., Murphy, E., & Phoenix, J. (2017). *Marginalization, young people in the south and east mediterranean, and policy: An analysis of young people's experiences of marginalization across six SEM countries, and guidelines for policy-makers.* (Power2Youth Working Paper No.35).

Deneulin, S. (2014). *Creating more just cities: The right to the city and capability approach combined.* Bath Papers in International Development and Wellbeing, No. 32, University of Bath, Centre for Development Studies (CDS).

Fainstien, S. S. (2010). *The just city.* Cornell University Press.

Lefebvre, H. (1996 [1968]). The right to the city. In H. Lefebvre (Ed.), *Writings on cities* (E. Kofman & E. Lebas, Trans.) (pp. 63–184). Blackwell.

Lopez-Fogues, A., & Cin, F. M. (2017). *Youth, gender and the capabilities approach to development: Rethinking opportunities and agency from a human development perspective.* Routledge.

Miller, B., & Nicholls, W. (2013). Social movements in urban society: The city as a space of politicization. *Urban Geography, 34*(4), 452–473.

Shlomo O. (2016). Between discrimination and stabilization: The exceptional governmentalities of East Jerusalem. *City, 20*(3), 428–440.

Sen, A. (1985). Well-Being, agency and freedom: The Dewey Lectures 1984. *The Journal of Philosophy*, *82*(4), 169.

Sen, A. (1992). *Inequality re-examined.* Clarendon Press.

Sen, A. (1999). *Development as freedom* (1st ed.). Knopf.

Shtern, M. (2018). Towards 'ethno-national peripheralisation'? Economic dependency amidst political resistance in Palestinian East Jerusalem. *Urban Studies, 56*(6), 1129–1147.

UN Habitat. (2011). *Cairo: A city in transition.* Cities and Citizens Series, Bridging the Urban Divide. The American University in Cairo, UN Human Settlements Programme, UN Habitat for a Better Urban Future.

UN Habitat. (2013). *Urbanization and urban risks in the Arab region.* 1st Arab Region Conference for Disaster Risk Reduction, pp. 19–21 March 2013 at Aqaba–Jordan.

Yiftachel, O., & Yacobi, H. (2006). Barriers, walls, and urban ethnocracy in jerusalem. In *City of collision: Jerusalem and the principles of conflict urbanism* (pp. 170–176). Birkhäuser–Publishers for Architecture.

CHAPTER 2

Youth in Arab Cities

Introduction

Cities are complex constellations that can be spaces of both opportunity and inequality. Big cities in particular can be sites of extreme inequalities, in some instances even showing higher levels of inequalities in comparison with national levels (Glaeser et al., 2009). There is a sharp rise in urbanization worldwide, with an estimated projection that by the next decade, over half the world population will be living in cities (UN, 2019). This explains why "The right to the city" has gained traction with policymakers, development organizations, and academics pushing for urban policies, that promote inclusion and social justice, and call for the conceptualization of the "right to the city" as part of the human rights parcel (Purcell, 2013). In Arab cities youth constitute a large proportion of the population, with young people ages 15–29 making up around 30% of the population, and those under 30 making up over 60% of the population (UNFPA, 2020). At the same time, youth are amongst the most vulnerable groups in these cities. This chapter will explore youth experiences in the four cities—Jerusalem, Cairo, Tunis and Amman. It will reflect on the social, economic and political context in the four cities, through the two frameworks of Lefebvre's (1968) "right to the city" and Sen's (1999, 1992, 1985) "capability approach".

R. A. Nuseibeh, *Urban Youth Unemployment, Marginalization and Politics in MENA*, Middle East Today,
https://doi.org/10.1007/978-3-031-15301-3_2

As the neoliberal political and economic ideology spread throughout the world including Arab cities, and was pushed and promoted by international organizations, such as the World Bank and the International Monetary Fund (IMF), inequalities and segregation intensified in Arab cities. Neoliberalism, is a term coined in 1938, by Ludwig von Mises and Friedrich Hayek, who saw social democracy, and the development of Britain's welfare state, as manifestations of a collectivism, hence on the same spectrum as Nazism and Communism, and so they proposed this ideology to free capital from government intervention (Monbiot, 2016). However, this ideology proposes the liberation of the few from government regulation, at the expense of the masses and the environment. It wasn't until the 1970s that this ideology was extensively imposed intensively on much of the world, often without democratic consent. After Margaret Thatcher and Ronald Reagan took power, neoliberal ideology was imposed and implemented through massive tax cuts for the rich, destruction of trade unions, privatization, outsourcing of public services and deregulation (Monbiot, 2016). Similar policies have swept throughout the Arab world, as well, increasing inequalities and hindering democracy, leading to the majority of people to not have much of a say in how their cities are shaped.

The framework of the "right to the city", which is not framed in the language of economics, commodification and use value, is more concerned with people's lives and their human rights in the spaces that they are inhabiting (Deneulin, 2014). The urban space is examined through this lens, by how much it enables its residents—and in our case here youth, to exercise their rights equally, and how much they are able to shape the space they are living in. In this book I will be combining this framework with the "capabilities approach" framework of development, which calls for a paradigm shift in development policy from focusing solely on economic development, through increasing human capital, to focusing on enhancing capabilities, so people are able to function in ways that are meaningful to them and also people are able to lead the life that they have reason to value. Hence, as Sen (1985) stresses, there is a focus on "agency freedom," and in particular people's empowerment to be able to shape their cities and lead lives that they have reason to value.

The "Right to the City" and "Capability Approach" Combined

One of the most influential advocates of the "right to the city" is Henri Lefebvre (1968). The city that Lefebvre referred to is not the contemporary city as we know it today, in terms of size and density, but rather an urban space (Mayer, 2012). He has used different terms to refer to the "right to the city", such as "the right to centrality," "the right to space" and "the right to difference" (Shmid, 2012). The "right to the city" is thus defined as the collective rights of the inhabitants of cities, including minorities and vulnerable groups, giving them the legitimacy to organize the space and inhabit it, and be fully self-determined. This right includes civil, political, economic, social, cultural and environmental rights to the space (UN Habitat, 2017).

Lefebvre, who was a Marxist, saw that cities were shaped by capitalism and hegemonic national political powers. He believed that under capitalism the space in the city was divided into segments, with groups of different social classes inhabiting different zones and receiving different services (Purcell, 2013), Lefebvre has argued that this "right to the city" or to urban space is both a "cry and a demand" of those who are marginalized and deprived of basic material and legal rights in the city, those who are deprived of their basic social, political and civic rights, those who are prosecuted based on their religion, ethnicity, race, gender and sexuality; those who are excluded from the benefits of urban life, and deprived from participation in the decision making processes that affect their lives in the city (Marcuse, 2012). So Lefebvre sets an argument for the need to radically restructure social, political and economic relations in the city, and rework the structure of liberal democratic citizenship (Purcell, 2002). What is more, Lefebvre recognized the need to expand the scope of people's "right to the city" beyond the state structure, because through the state structure people are limited as to how much they have a say in the production of the urban space (ibid.). The "right to the city" calls for the expansion of the rights of urban inhabitants beyond the limited structures of the conventional citizen enfranchisement and gives them a direct voice in any decision that contributes to the production of their living space (ibid.). Under the "right to the city", Lefebvre calls for the shifting of power from capital and hegemonic national political powers, towards urban inhabitants who have earned the right to the space by living out the routines of everyday life in the city (ibid.). Therefore, in the long run, winning the "right to the city" for all may be a win–win for all those residing in the city, but in the shorter run it will involve conflict, as it will entail restructuring of the political system to eliminate the rights to dispossess others, to exploit, to dominate and to suppress others (Marcuse, 2012).

The idea started to materialize in the global arena in 2011, with the "Global Charter for Human Rights in the City," which aims to promote and strengthen the human rights of all inhabitants of all the cities in the world (United Cities and Local Governments, 2012). The charter emphasized the democratic aspect of the "right to the city" stating that "All city inhabitants have the right to participate in political and city management processes, in particular, to participate in the decision-making processes of local public policies, to question local authorities regarding their public policies, and to assess them, and to live in a city that guarantees public transparency and accountability" (United Cities and Local Governments, 2012: 11).

An additional framework for the "right to the city" has been offered by Jabareen (2014, by analyzing and merging the different concepts of the "right to the city" offered in Lefebvre's writings. He proposes six major rights to explore: the right of appropriation, the right of participation, the right to centrality, the right to inhabit, the right to habitat and the right to individualization in socialization. Analyzing Lefebvre's work, Jabareen (2014: 136–137) defines the "right to appropriation" as the right of complete usage of the urban space, which entails the right to use the space; to live, work, play, study, etc. in the space. The "right to participation" includes people's involvement in social, cultural, spatial and economic space production, and to participate centrally in any decision that contributes to the production of urban space. The "right to centrality" opens the urban space for people to use the center, and not be condemned to marginalized peripheral spaces. The "right to inhabit" includes giving power and voice to those inhabiting the city. The "right to habitat" includes decent housing, safety, a clean environment and basic infrastructure, and finally the "right to individualization" in cities, which includes individual freedoms.

In Middle Eastern cities, policies of privatization, crony capitalism, and neglect of social programs and welfare provision have intensified urban inequalities; leaving large numbers of inhabitants vulnerable to inadequate infrastructure, pollution, poor service delivery and lack of job opportunities, isolation and lack of influence over decision-making institutions in the city (Bayat & Biekart, 2009). These urban inequalities in Arab cities have pushed people to ask for social justice, freedom, accountability and democratization, which has caused a surge in the "right to the city" movements in Arab countries. However, these movements according to Al-Hamarneh (2019) are still weak, as people are struggling for basic democratic rights, freedom of speech, human rights and political participation. So for youth to be able to reclaim their cities, they need to have what Amartya Sen (1985) referred to as "agency freedom." Here comes the framework of the "capabilities approach." Sen (1999) offers an alternative lens to development, which is the "capabilities approach", which seeks to provide the capabilities necessary for the traditionally excluded to envision alternate choices (Adely, 2009). Sen's (1999) approach assesses people's well-being in terms of their ability to function, and whether they are provided with the capabilities to function in ways that matter to them, so that they can choose the lives they have reason to value (Lopez Fogues & Melis Cin, 2017). Capabilities are the opportunities and the freedoms people need, in order to achieve the various lifestyles they desire, and as a result live the life that they have reason to value (Anand et al., 2005). Ginwright and Cammarota (2002) have suggested that in the field of youth development the focus should be on examining larger socio-political and economic factors that contribute to everyday youth struggles and harm their psychological, mental and spiritual well-being. The choices of youth are constrained by the social, political and economic resources available to them, therefore examining the opportunities available to them helps in understanding youth exclusion in Arab cities.

Urban Segregation and Social Exclusion

Cities bring together people with diverse backgrounds, but this can result in sub-divisions of spatial segregation as people sharing common characteristics often congregate in their own neighborhoods within the city. This intra-urban segregation has meant that the most vulnerable social groups often find themselves in the most deprived neighborhoods in the city. Spatial segregation can be a result of the differential financial capacities of people to choose their place of residence, it can also be a result of exclusionary municipal policies, or of indirect community exclusionary measures, where certain groups are denied access to affluent neighborhoods, based on their ethnicity, religion, social class, sexual orientation, family structure or social status. As a result of these exclusionary measures the most socially vulnerable groups often find themselves confined to deprived neighborhoods with high levels of crime, weak accessibility, low housing prices, predatory loan schemes, low-quality municipal services, including low educational services and infrastructure. The relationship between spatial segregation and educational and economic outcomes can be a story of a vicious circle, where sustained exposure to disadvantage leads to segregation, and segregation leads to more inequality and disadvantage (Van Ham et al., 2018). Socialization effects in deprived neighborhoods, such as lack of positive role models, lack of social networks, negative peer groups effects and stigma, can also affect the employment prospects of residents of those neighborhoods, which further limits intergenerational social mobility (Van Ham et al., 2013). Children born in deprived neighborhoods often find themselves raising their families as adults in the same neighborhoods.

Urbanization trends in Arab countries have come with persistent problems of intra-urban segregation, such as the presence of gated communities for the wealthy and extremely impoverished neighborhoods with weak infrastructures, low-quality services and high levels of pollution for the urban poor. The urban restructuring occurring in cities like Cairo and Amman; brought about by the competition to attract international investments, together with the internationalization of commercial real estate, and construction companies capable of providing high-quality services, has created those "urban islands", that cater to a lifestyle of excessive consumption for the wealthiest niche, while the most vulnerable social groups are condemned to deprived neighborhoods (Daher, 2013).

> People are not used to life outside the compound. They have everything in the compound; hairdresser, doctor, anything you want is inside the compound. Once they leave the compound it's a different world. They start facing reality. It's a completely different city. They don't recognize the people outside the compound.[1]

> I'm living in a very undeveloped area and every day I commute from this undeveloped area to the American University of Cairo and it's like going from hell to heaven....Here we have pollution everywhere, we don't have rubbish collectors, we literally throw the rubbish on the street. There's a

[1] Interviewee Haya (Pseudonym), 19, Christian, Female, Economics Student, Cairo.

> street for garbage, what can we do. When I go to the campus, it's so clean, we have so many trees and nobody is allowed to smoke there.[2]

Examining Arab cities through the lens of Lefebvre's "right to the city" is complex. Morange and Spire (2019) have noted that Lefebvre's concept of the "right to the city" has become the subject of multiple re-appropriations by mostly anglophone neo-Marxist geographers of the Global North, who have taken this concept that was born in Fordism Europe and revisited it to look into the effects of privatization policies on European and North American cities. They note that the adaptation of these concepts to the Global South is more complex. The history of post-colonialism in the MENA region and continued "urban colonialism" of East Jerusalem (Shlomo, 2016) means that the "right to the city" debate is being visited here in a more complex landscape. Additionally, there are several elements at play in the Arab communities, such as traditional tribal authorities, religion, customs, informal labour exchanges, refugee crises and the contemporary surge in online activism, that all influence the landscape in the Arab city.

Another phenomenon present in Arab cities is the formation of informal haphazard neighborhoods formulated by squatters. Séjourné (2009) argues that this phenomenon is caused by a combination of rapid urbanizations, combined with cuts in public expenditure, and privatization policies. Cairo has four out of the thirty biggest slums in the world, which are called in Egypt *Ashwa'iyyat* meaning haphazard; these are Imbaba, Ezbet El-Haggana, City of the Dead and Mansheiet Nasser (Khalifa, 2011). The *Ashwa'iyyat* suffer from problems of accessibility, dilapidated streets and infrastructure, high residential densities and lack of services. Although in Cairo there is no geographical segregation between the rich and the poor, as poor people can be found living in urban pockets in middle-class areas, the highest concentration of the most vulnerable social groups can be found in the *Ashwa'iyyat*. Cairo is one of the largest cities in Africa and the Arab world, with an estimated population of 21.3 million (UN, 2018). It is one of the ten most populated cities in the world, and it is undergoing the hard experience of hyper-urbanization (UN habitat, 2011). Greater Cairo (GC) today is made up of five governorates, Cairo, Giza, Sixth of October, Qalubiyah and Helwan. There is no administrative macro-level body in charge of Greater Cairo as a

[2] Interviewee Salim (Pseudonym), 19, Muslim, Male, Political Science Student, Cairo.

single administrative entity, each governorate has its own administrative structure (Schechla, 2015). The boundaries of the city are unclear, and constantly changing, which creates confusion in terms of statistical analysis, and issues of jurisdiction. Some say that 65% of Cairo's population are living in informal slum areas (Séjourné, 2009).

Although the development of the *Ashwa'iyyat* has started with rural migration to the city and inept planning, other factors are still contributing to this phenomena today, including population growth, inadequate and corrupt investment practices in the housing sector, and the adoption of economic liberalization policies and reduction of social spending in the state budget (Schechla, 2015). The private sector dominates the total formal real estate supply at 80%, so while luxury real estate development has remained a lucrative business, it only serves an estimated 3% of the population (ibid.). In fact the city is said to be filled with buildings that are empty (an estimate of 7.7 million vacant units), but these buildings are inaccessible to impoverished Egyptians. This means that the problem of housing is not primarily caused by housing shortage, but by housing maldistribution. Most of the formal urban advances exemplified by New Cairo City, for example, Qattameya, Mena Garden City and Sheikh Zayed City, to the east, and Dreamland and Beverly Hills, to the west of Giza, cater to the rich elite Egyptians (ibid.). At the same time around a third of the Egyptian population is living under the poverty line (CAPMAS, 2021). In Cairo the picture is grimmer, with official data severely underestimating the real figure, due to the fact that large numbers of the city residents are not even counted in official census, as they reside in informal settlements*Ashwa'yyat*. Where you reside in the city can also determine your life opportunities. Being born and residing in certain neighborhoods in Cairo determines people's life prospects and the social services available to them. Income inequality in Cairo is higher than the national level. The Gini coefficient in Cairo is 0.4 in comparison with the national level of just over 0.3 (with Zero indicating perfect equality and one indicating perfect inequality) (Bourkinas, 2021).

Critics of the current government's urban policies have pointed out that instead of following inclusive growth policies and investing adequately in public infrastructure, the Sisi government is investing billions to create a New Administrative Capital (Loewert & Steiner, 2019; Mandour, 2021). The New Capital is envisioned to house around 6 million people, built in the desert—40 km east of Cairo, and it is intended to host a new government quarter, including the parliament, ministries, the supreme council, diplomatic missions and a new presidential palace, skyscrapers, an airport, and an amusement park (Loewert & Steiner, 2019). The idea of the new city was promoted to investors as a "Smart City" that will ease congestion in Cairo and is being constructed to serve all, "*Madinat al-Jami,*" a "City for Everyone" (ibid.). However, as the urban professor Galila Al Kadi explains, since the times of the pharaohs "administrative capitals were established for the purpose of isolating the ruler from the people" (AFP, 2021).

Critics of the project were fast to point out that the prices of the flats in the new city reflect that it is out of reach for most Egyptians, and is only available for Egyptian elites, who are benefiting from the current government's policies, and are less likely to violently revolt. The New Administrative City seems to be adding to segregation and inequality in Cairo, by shifting the urban center of the city, and surrounding this new center with political elite and supporters of the regime, hence limiting any urban unrest and dissent.

Mandour (2021) points out that the construction of the New Administrative Capital coincidess with a concentrated effort to alter the urban landscape of Cairo, by shifting the political and economic weight eastward away from the current center, where segments of poor populations reside and can cause social unrest; building bridges on the east side to allow for deployment of military forces in case of social upheavals, as well as "reengineering" the social composition of slum areas close to government quarters, by vacating slums and relocating residents to the outskirts and replacing them by upper-middle-class residents.

The idea of building new cities with their own economic base in Cairo was suggested as early as the 1950s, but the actual construction of those new cities started during the Sadat era in the 1970s. The New Urban Communities Authority (NUCA) was founded in 1979 as the official governmental authority in charge of overseeing their development (Tadamun, 2015). The mission of this authority was to use the empty desert surrounding Cairo to absorb the demographic expansion, decrease the expansion of informal settlements and alleviate the congestion in the city. However, Tadamun (2015) argues that the NUCA has consistently deviated from its social role of serving the people, to embracing the role of a real estate developer. Centralization in Egypt makes it also very difficult for the community and local actors to actually influence policy, which has led to the creation of "ghost cities," (Tadamun, 2015) where billions of government funds were poured, but are mostly empty, instead of serving the people living in the urban core of Cairo. Sims (2014), in his book on Cairo's new cities, examines 20 of those new cities, built since the time of Sadat and he says that these cities were designed without taking into consideration the needs of the majority of Cairenes. They were too big, expensive and far away from the heart of Cairo, and therefore one needs a car to drive to those cities, they also do not allow shops to open in the new buildings, like kiosks and small businesses, where most Cairenes people create an income. That's why by 2014, these new cities housed less than a million Cairenes. These cities could not become self-sustained economically, they also suffered from lack of social services such as schools, clinics and centers and lack of transportation and connection to the city center where economic opportunities are, so people were/are reluctant to move into those new cities and prefer to remain in the congested centers, where economic opportunities exist (Sims,2014).

Tunis has a problem of informal neighborhoods as well, which they refer to as *gourbiville* (Stambouli, 1996). The first *gourbivilles* developed in Tunisia in the 1930s and 1940s. People who had lost their livelihoods as a result of their uprooting during the process of land confiscation under French colonial rule started migrating to the city. They began to build precarious housing and settlements made from gourbiville (mud) or bidonville (tin), hence the names of these slums (Stambouli, 1996). Their inhabitants squatted on public land without access to basic urban services, such as water supplies, sewers, electricity and roads (ibid.). Between 1960 and 1970, the state refused to give legitimacy to these haphazard settlements and launched a program to demolish them, and evict their residents. However, their endeavors were unsuccessful as people only moved to bigger and more overcrowded haphazard settlements, that were harder to demolish (Chabbi, 2010). In 1978, massive demonstrations took place by residents of those neighborhoods, who continued to live in unsanitary, and crowded conditions; the demonstrations have resulted in the death and injury of several demonstrators at the hands of state police (ibid.). These events led the government to pay attention to these deprived neighborhoods, and in 1981, the Urban Renovation Agency was created, to renovate and equip the *gourbiville*, which unintentionally led to their expansion (ibid.). Today the city's housing structure is divided into three poles; in the north, the wealthy populations enjoy luxuriously designed and well-serviced residential areas, in the south, the housing for lower-income groups stretches amidst polluting industries, and in the west of the city, spontaneous and informal *gourbiville* (Stambouli, 1996). The life opportunities for youth are directly influenced by the environment they are living in (Brooks-Gunn et al., 1993). Residing in these impoverished settings can result in the denial of a person's capability to lead a healthy life; they are denied basic social services and most problematic is their limited ability to have access to quality education and have a relevant and to realize their full potential.

> There are areas that are known in Tunis that the police do not enter, so you can imagine the situation there. What do we expect from children being raised in that environment?[3]

To be able to rent or buy property in the more affluent areas in Tunis, means that people need to have access to wealth or a steady income covering the rent cost, which is more problematic for youth who suffer from higher levels of unemployment and precarious working conditions.

> People can't rent on the salaries they are getting. They will need to find an apartment far away and count on public transportation and buses which get delayed and canceled a lot.[4]

Amman has gone through urban restructuring over the past twenty years, where the spatially engineered reality has created high-end business towers, such as Abdali, upper-end residential gated communities, such as green land, and low-income residential cities on the periphery, with the aim to push the poorer segments of society to the outskirts of the city in newly zoned "heterotopias" (Daher, 2013: 101). Groups from the lower socio-economic status in Amman were pushed outside the center into low income housing in Jizza, Abu-Alanda, Marka and Sahab, which lack basic services, such as solid waste management, water infrastructure, good quality roads and accessibility to schooling (ibid.). Meanwhile, the upper-end western gated communities in the city consume high amounts of water resources for luxurious swimming pools and landscaping, with disregard to the water scarcity in the country (ibid.). In their drive towards "modernizing" Amman, a country like Jordan finds itself shifting its priorities from the provision of crucial social services, such as quality education, health care, social security and social housing, to indirectly

[3] Interviewee Lima (Pseudonym), 25, Muslim, Female, NGO Employee, Tunis.

[4] Interviewee Khalid (Pseudonym), 23, Muslim, Male, Unemployed, Tunis.

subsidizing and facilitating the work of multinational corporations (ibid.). These multinational corporations start to have more of a say in how the city looks like than its actual inhabitants, which curtails democracy and further fuels spatial segregation and inequality. Informal settlements are less present in Amman, in comparison with Cairo and Tunis. However, they are on the rise, and it is estimated that they are rising at a rate of 4.3% per year, especially in Greater Amman area (Abed et al., 2015). There are socio-economic as well as political factors behind this steady increase in informal settlements (ibid). The high cost of properties in Amman, high interest mortgages, as well as poor job opportunities, make it hard for people to buy formally in the city (ibid). There is also informal building activity around refugee camps, with UN data revealing informal activity and expansion around camps. The influx of refugees from neighboring countries, and the internal rural/urban migration of those seeking job opportunities, has also increased housing demands, pushing people towards informality in housing (Abed et al., 2015).

As for the case of Jerusalem, it is rarely grouped with other Arab cities in research, given its contested sovereignty issues. Today under international law, Jerusalem is divided territory with its western/southern side considered Israeli territory and eastern/northern side considered Occupied Palestinian Territory and referred to as "East Jerusalem" (Ben-Hillel, 2013). Israel's capture of West Jerusalem in 1948 went unchallenged by the international community, despite its illegality, as Jerusalem was recognized as a corpus separatum (Schechla, 2014). This has resulted in the dispossession and expulsion of 39 Palestinian communities consisting of 97,949 people from West Jerusalem, and the confiscation of Palestinian land (ibid). The expulsion of Palestinians from the city continued after the occupation of the eastern side of the city in 1967. Since the occupation, Israel has worked on changing the urban geography of the city (Asali Nuseibeh, 2015, 2019). It has expanded the municipal borders of the city by including within them Israeli Jewish settlements and cutting out Palestinian populated neighborhoods, by installing the Separation Barrier to control the demography of the city and access to the city. The state also limits Palestinian expansion in the city, by the restrictions on issuing building permits and neglecting the urban planning and expansion of Palestinian neighborhoods (Ir Amim, 2021). This political and spatial expansion over Palestinian spaces in Jerusalem has been described as a type of "colonial urbanism," in which Israel has engineered a structured ethnic expansion over Palestinian spaces and has suppressed the

Palestinian community, in order to control its "unified" capital (Shlomo, 2016 cited in Asali Nuseibeh, 2019). Shlomo (2016) explains that while Israel has claimed sovereignty over East Jerusalem, it has never administered the eastern side of the city the way it did other Israeli cities, due to a combination of discrimination, unwillingness and inability, on the one hand, and Palestinian resistance, self-exclusion and separatism from the Israeli State apparatuses, on the other (cited in Asali Nuseibeh, 2019). Shlomo (2017) argues that Israel's urban governance of East Jerusalem lies outside Israeli governmental and administrative norms and, in some cases, even outside Israeli law. He explains that the Occupation of East Jerusalem is not carried out by the military, as is the case in the West Bank, instead it is carried out by state and municipal civil apparatuses. Israel has a dilemma when it comes to the Palestinian population of East Jerusalem. It considers the land Israeli territory, but it considers the Palestinian population, which constitutes today around 40% of the population of Jerusalem, as residents and not as citizens of the state. The legal status of Palestinian Jerusalemites in the state of Israel reflects the way the state views their rights as residents, which is conditional and devalued. Since the beginning of the occupation, the state has treated them as new immigrants rather than as citizens who have the right to live in the city (B'Tselem, 2017 Asali Nuseibeh, 2019). It has granted them residency permits, which grant them health insurance and social security benefits, the ability to travel throughout Israel, and to live and work anywhere in Jerusalem, and eligibility to vote in municipal elections (Jonathan et al., 2018). However, they cannot participate in any political process such as voting or standing for elections in the Knesset (Parliament). They do not have equal civil, political and social rights with Israeli citizens (see Asali Nuseibeh, 2015). If they reside outside the Israeli municipal limits of the city, even in neighboring cities in the West Bank, for over seven years, they lose their right to this residency permit. If they marry a person from the West Bank and Gaza, they cannot do family unification and live legally in Jerusalem as married couple, because Israel denies Palestinians from the Occupied Palestinian Territories to be granted Israeli citizenship or residency by marriage (Haaretz, 2022). Even the so-called "permanent residence" for Palestinian inhabitants of Jerusalem is not necessarily permanent.

> Having a residency permit in Jerusalem is a problem in itself. It's like a risk always, that you might lose it anytime. For example, if I want to get

> married, it has to be with someone from Jerusalem, because I have to think of the children and what documents they will have, and I have to think where my center of life is. There are all these dilemmas that one needs to think of, so you are never really feeling stable.[5]

> My father has been waiting for 26 years to get his Jerusalem ID. Every now and then we get an appointment with the ministry of interior and they just renew his temporary permit to stay in Jerusalem, but they don't grant him the unification and they don't give him an ID.[6]

The amendment to citizenship law prevents the marriage unification between Palestinians holding Israeli citizenship and those holding residency permits with Palestinian family members from Gaza and the West bank. The Palestinian family unification ban was approved in the Israeli Knesset this year 2022. The Israeli Minister of Interior in an interview with the Israeli newspaper Yedioth admitted that the reason behind this policy is to prevent the "creeping right of return," of Palestinians (Haaretz, 2022).

The urban segregation in Jerusalem can be seen along ethnic lines. The Association of Civil Rights in Israel annually publishes figures reflecting the discrimination that Palestinian neighborhoods face in Jerusalem. Only 8.5% of Jerusalem as a whole is allocated for residential use by Palestinian residents, although they represent about 40% of the city inhabitants (ACRI, 2021). As a result, the density of housing in the Palestinian neighborhoods of East Jerusalem is almost twice as much as that in the Jewish neighborhoods in the west side of the city. Due to the Israeli restrictions, many Palestinians had to resort to building without getting permits from the municipality, which then results in an increase in the rate of demolitions in Palestinian neighborhoods (ibid.). Among other political and economic factors, the huge gap between the municipal services offered to Palestinian and Israeli neighborhoods has limited Palestinian opportunities and life prospects, leading to reduced quality of life, in comparison with West Jerusalem. In a report by UN Habitat and the League of Arab States (2016) they look into "urbanization under occupation" in East Jerusalem, and they acknowledge how the Israeli Occupation policies have negative repercussions on the Palestinians ability to develop long-term plans and strategies for sustainable housing and development. The

[5] Interviewee Nadim (Pseudonym), 29, Christian, Male, Jerusalem.

[6] Interviewee Taline (Pseudonym), 24, Muslim, Female, Jerusalem.

report shows how Israel does that through several policies; expansions of Israeli illegal settlement on Palestinian land, restricting and limiting Palestinian expansion through bureaucratic processes, erecting a separation barrier that infringes on Palestinian land and stops the natural expansion of Palestinian communities and separates between Palestinian communities, and the deliberate and systematic destruction of Palestinian infrastructure and houses (UN Habitat, the League of Arab States, 2016). Jerusalem is the opposite of Cairo, Amman and Tunis in that regard; there are no informal settlements, urban expansions are highly monitored by the Israeli State and Palestinian expansion is fought fiercely (ICAHD, 2022). In the next section we will look at how our four cities are governed and what leads to such huge disparities and segregation in them.

Municipalities and Governance

Contrary to orientalist scholarship on Islamic cities, that say that municipalities were European concepts, which were alien constructs in the Middle East, Ottoman municipalities were established across Arab provinces—first in Tunis (1858) and Beirut (1861), then Jerusalem (1863), Tripoli, Libya (1867), Nablus, Baghdad and Damascus (1868) (Hanssen, 2018). These institutions were created as a response to urban crises, such as inadequate provision of public services, particularly health care, and as a result of economic competition between Arab-Ottoman merchants and foreign rivals (ibid.). Although local governance is a democratic concept that gives power to inhabitants of cities, the first municipal elections did not constitute universal suffrage and they were influenced by exclusions based on class, and gender. They were subsequently also undermined by colonial powers later, as they appointed state officials and bureaucrats, rather than holding democratic elections.

Although in the last twenty years there has been a drive towards decentralization in the Arab world, granting municipalities more power and more of a say in shaping policy, municipalities are still limited in the scope of their authority and are mostly controlled by higher levels of government. Municipal governments are supposed to be autonomous authorities, independent from the central government and elected directly by the people, with a mandate to develop their territory and implement their programs for their cities. It has been argued that this process of granting municipalities autonomy from the central government has several positive effects, such as reducing corruption, increasing transparency and accountability, increasing efficiency, and giving the inhabitants of cities more of a say in the decision-making processes (Yerkes & Muasher, 2018). Local officials who live among the people can better identify their pressing needs, and so this leads to better planning and provision of services, which in turn leads to better tax revenues (Yerkes & Muasher, 2018). For example, in Porto Alegre, Brazil, decentralization has led to better municipal services and the doubling of elementary and secondary school enrollment between 1989 and 1996, the city has also increased its revenue collection by 48% (ibid.). Decentralization makes it also easier to implement entrepreneurial and creative changes at the local level, more so than at the federal level as it requires less resources, and it is also easier for local officials to take risks, in comparison with national level bureaucrats (ibid.)

In Amman, in 2015, on the legislative level, decentralization law and municipal law were enacted, with the objective of increasing public participation in decision making at the local level (World Bank, 2017). The idea behind the enactment of these two laws was to expand participation in democratic elections at the local level, strengthen the independence of municipalities and increase their responsibilities, in the hopes of addressing people's needs and improving service delivery (ibid.). The country held municipal and governorate council elections in 2017, in a bid to give voice to marginalized groups in shaping their cities. Jordan is divided into 12 governorates, each of them is headed by a governor appointed by the king, and under each governor there is a non-elected executive council and a previously non-elected consultative council, whose role is to advise the executive councils on local needs (Ranko et al., 2015). With the new system of decentralization, the executive councils are to be elected, or at least 85% of their seats are to be elected (ibid.). The role of these councils is to present recommendations on development projects in each of their localities, although their recommendations are non-binding (ibid.). The 12 governorates comprise 99 municipalities, plus the Greater Amman Municipality, that are independent entities under the supervision of the Ministry of Municipal Affairs (MoMA) (ibid). However, the Greater Municipality of Amman, which is the largest local government in Jordan, reports directly to the prime minister and its mayor is directly appointed by the prime minister (World Bank, 2017). The metropolitan area of Amman contains around 4.6 million- 42% of the country's population (Department of Statistics, 2021). Amman is also increasingly hosting citizens from neighboring countries, fleeing conflicts such as in Iraq, Syria and Palestine.

While elections at the municipal level have existed since 1925, the roles of municipal council have, so far, been minimal to minor development projects and street infrastructure (Ranko et al., 2015). The limited role of municipality councils in combination with high levels of tribal affiliations in Jordan and weak political parties has led to political apathy and lack of participation and political activism. Therefore, when Jordan held its municipal and governorate council elections in 2017, the voters turnout was very low. Only thirty percent of eligible voters turned out, with an especially low turnout in Amman in particular (16%), which indicated that people did not believe that the government was serious about democratic reform (Al khalidi, 2017, Ranko et al., 2015). Opposition parties in Amman maintain that there's a gerrymandering problem, where an electoral law magnifies the votes of sparsely populated tribal areas, which mainly support the monarchy, at the expense of larger cities where the opposition has more support (ibid). For example, in the governorate of Amman, where the Palestinian population mostly resides, one parliamentarian seat represents 96,000 people, whereas in Tafilich and Ma'an, that are mostly populated by East Bankers/Jordanians and those loyal to the Jordanian monarchy, each seat represents between 22, 000 and 30, 000 residents (Muasher, 2011 cited in Harris, 2015).

In Tunisia, reforms towards decentralization acquired momentum after the 2011 revolution. Before the revolution Tunisia's political system was highly centralized and only served the interests of the pro-regime elite. During the governments of both Bourguiba and Ben Ali, local officials were handpicked and security concerns were the major issue in their appointment; municipalities were highly dependent on the central government and their decision-making processes were limited to menial issues, such as waste collection (Yerkes & Muasher, 2018). They did not have a say in vital service provisions that the city inhabitants needed, such as education and health care (Yerkes & Muasher, 2018). Huge disparities existed between municipalities in Tunisia, with eighteen municipalities, among them the city of Tunis, awarded 51% percent of the state's municipal budget in 2013, while the other 246 municipalities had to make do with the rest (Bellamine, 2015). Therefore, Tunisia's decentralization efforts were aimed at reducing these disparities between the different regions and to empower local actors to lead the development initiatives in their communities based on their communities' needs (ibid.) (2018. The first local elections in Tunisia took place in 2018, ten days after the

Local Authorities Code—which is the law governing the entire decentralization process—passed by parliament (ibid). Although the elections were democratic and fair, the turnout was low at 34% (Council of Europe, 2018). Explanations for this low voter turnout included, people's lack of confidence in the electoral system and the political class, as well as lack of awareness of the importance of decentralization (bid.). Another contributing reason could have been the low media coverage of the campaign, due to strict rules on the allocation of air-time to political participants (ibid.). However, traditionally excluded groups did manage to secure some successes in the elections. Tunis elected its first female mayor Souad Abderrahim from Ennahda party, and 47% of the elected officials were women and 37% were youth (Carniege, 2018).

Cairo did not achieve the same democratic transition as Tunis after its revolution. Egypt has 27 governorates, each of is subdivided into a region, a capital city, and a group of smaller cities or villages, and each city is made up of constituent neighborhoods (Schechla, 2014). In Egypt, there are no elections for local councils and mayors. The prime minister appoints the heads of regions and military men are appointed by the executive branch of the government to run governorates and municipalities. Neither the constitution nor governmental practices in Egypt have ever given any power or autonomy to local governments (ibid.). Egypt does not allow any local government the power to set its own agenda, raise and spend its own revenue, or make decisions about development and answer people's demands (Tadamun, 2013). It has only local administrations, mostly appointed by the central government and serving the national rather than the local agenda, which means that their policy and development projects are mostly one-size-fits-all and are not customized to address constituents' priorities (ibid.). The constituents are also deprived of their right to elect officials, which are to be held accountable only to them. In 2014, Egyptians approved a new constitution, which gives local administrations more financial independence, but in practice local governance still remains centralized. Municipalities in Egypt are dependent on the central government for funding, which is very meagre and it stands at only 11% of the state budget, less than half of the municipal budgets in most emerging economies (Schechla, 2014). The Egyptian local communities have no authority or a legal framework to question this poor funding and lack of autonomy; there is no accountability or transparency in allocation of resources. The situation is even more elusive and less participatory in Cairo's slums, where there is even less money to go around, as the number of inhabitants are severely undercalculated (Schechla, 2014). This undercounting affects budgetary decisions, and provision of public services, such as schooling and health care, causing overcrowding and strain on the services provided.

As for Jerusalem with its two distinct "sides", the eastern occupied Palestinian side and western Israeli side, the municipal story is more complex. The destruction of urban democracy in Jerusalem goes back to the British colonial power (the British Mandate), which employed a strategy of de-municipalizing Ottoman Jerusalem. They replaced municipal electoral processes with government appointments, reduced the council's Muslim membership, and established a sectarian quota, which lasted until 1948 (Hanssen, 2018). When the Jordanian authorities took charge of Jerusalem in 1948, government bureaucracies were shifted to Amman. The Palestinian side of the city suffered as a result of the impenetrable border between the western and eastern sides; it was cut off from water sources, from the ports on the Mediterranean sea, and the city growth stagnated (Dumper, 1997). During the Jordanian rule, municipal officials in East Jerusalem were complaining about the neglect of different parts of the city, especially Silwan and Shiekh Jarrah (Hanssen, 2018). Two decades later, urban democracy in East Jerusalem was destroyed further. In a breach of international law, Israel annexed East Jerusalem and expanded the city border to incorporate 28 surrounding Palestinian villages (B'Tselem, 2017). It extended Israeli law over East Jerusalem and started populating it with Israeli settlers, which was seen as a move to prevent any future separation of the city or the realization of a Palestinian state, with East Jerusalem as its capital. Altogether, there are 358,804 Palestinians living in East Jerusalem and they constitute around 38% of the residents of the whole city (ACRI, 2021). However, East Jerusalem is also inhabited by around 209,000 Israeli settlers, who, therefore, constitute around 40% of the Eastern side of the city (ACRI, 2021). The western side of the city is predominantly Israeli Jewish at 98% (ibid.). Since 1967 Palestinian East Jerusalem has not had its own municipality and has been governed by the Israeli Jerusalem municipality. Palestinian East Jerusalemites have boycotted the elections as a rejection of Israel's sovereignty over the city, which has led to their being disenfranchised for decades (Asali Nuseibeh, 2019). The Palestinian Authority appoints a minister for Jerusalem, but her/his authority is limited, and her/his role remains symbolic.

Israel holds local elections for mayors and councils every five years, and while the mayor is chosen by direct vote, the city council is chosen by party vote. The last time Israeli voters went to polls was in 2018, with the overall national turnout in local elections reaching 59.5%, and in some cities as high as 66.26% (Israel Ministry of Interior, 2018). However, the turnout of eligible voters in Jerusalem was not as high, and in the last elections the turnout was 39.29% (ibid.). The main reason for this low turnout is Palestinian abstention, although Palestinian citizens of the state of Israel have a high level of turnout in local elections, even higher than their turn out for national elections (Halabi, 2014). Palestinian East Jerusalemites are getting increasingly frustrated with their lack of power over policy and resource allocation to their neighborhoods. There is a growing feeling of despair as a result of their inability to build and expand their neighborhoods, their difficult economic situation, problems with their residency rights and ability to move and work freely, in addition to the checkpoints surrounding the city. There are also high levels of frustration from the weak Palestinian leadership. There is a feeling of despair among Palestinian Jerusalemites. On the one hand, there is a deep sense of political fatigue and disillusionment, and on the other, there are the economic concerns. Therefore, Palestinians are leaning towards fighting for their rights within the Israeli system, by both applying for Israeli citizenship and in a new shift by voting in the Jerusalem municipal elections. A 2018 poll found that 58% of Palestinian Jerusalemites supported the idea of voting in the city's elections (Blake et al., 2018). If Palestinian residents of Jerusalem (who constitute close to half the population of the city) voted at the same rate as Israelis, they could gain substantial representation in the city hall (Blake et al., 2018). In the last Jerusalem municipal elections, the Arab Joint List party attempted to enter the city council elections, but they didn't obtain enough votes. However, the very fact that they entered the institutional processes of the city council elections indicates that there is an interest in ending the boycott of the elections (Nikolenyi, 2020). The internal Israeli political fight in Jerusalem's local elections is fierce. The composition of Israeli society in Jerusalem is different from that in other large Israeli cities, like Tel Aviv and Haifa. While secular Jews account for a majority of the Jewish population in Tel Aviv and Haifa, the corresponding share of secular Jews in Jerusalem is only 21% (Nikolenyi, 2020). Conversely, whereas ultra-orthodox (*Haredi*) Jews make up only a very small minority of the residents in Tel Aviv and Haifa, they constitute 34% of the Jerusalem population (ibid.). Therefore, the dominant position of the Jewish ultra-orthodox parties remained unchallenged in the city council elections (ibid.). The mayor who won the Jerusalem elections of 2018 was Moshe Lion who received the support of three political parties (Yisrael Beitenu, Shas and Degel HaTorah), which has placed him in the camp of the conservative and ultra-orthodox religious communities, and this means that the city council of Jerusalem continues to cater to the residential needs of the ultra-orthodox community (ibid.).

Youth and the City

Youth exclusion has been very much debated especially after the 2011 uprisings. Calder et al. (2017) have found the very term "exclusion" problematic, because it implies that to be "included" is the ideal outcome. They explain that for Arab youth residing in authoritarian and patriarchal states, inclusion in society can mean disempowerment rather than empowerment. Sika (2016) has also noted that in the case of Egypt only young women and men who accepted the authoritarian system were included in the small venues of formal participation in public life. Calder et al. (2017) argue that in society people do not suffer from total exclusion, people can be included in one domain and excluded in another; therefore, they suggest that there is a need for a more subtle way to understand power relations within societies. When looking at youth in the Middle East, it is easy to put them together in one segment or look at them the way Cote (2014) suggested as a social class. France and Threadhold (2016) agree with Cote that there is a general process of proletarianization of youth in the last few decades, with youth suffering disproportionally from unfair wealth distribution. However, they disagree with looking at youth as a "one-lumped class" (France & Threadhold, 2016: 612). They recognize that the levels and severity of youth exclusion vary along intersectional lines of gender, sexuality, ethnicity, religion, sect and social class (ibid.). These different socially recognized identities create boundaries between groups of people in society, which can range between being purely symbolic, to becoming institutional and structurally enforced by law, which results in unequal access to resources and social opportunities (Lamont & Molnar, 2002). Therefore, I will be looking here at power distribution in cities, and individual freedoms that the youth have, in order to make the choices that they value, based on Sen's (1999) "capability approach", rather than just outcomes, such as employment and income.

> Most youth, you never find any of them doing what they love. For example, I would love to be an artist or anything, honestly you won't find anyone encouraging you. Even the country doesn't encourage you. This is the problem nobody is doing what they love.[7]

There is a need for reframing the debate on disadvantaged youth, by focusing on what informs what they value, their ability to choose what they value, and their access to the resources they need to achieve what they value, rather than just focusing on the material outcomes, whether they have an income or a career (Egdell & McQuaid, 2014).

Arab youth are not only economically excluded; they face exclusion along a number of interrelated dimensions, preventing them from participating fully in the urban space. These interrelated dimensions can include one or more of the following: the exclusion from housing, education, healthcare citizenship, political participation, employment, financial and physical assets, and access to credit (Chaaban, 2009).

> Many things disappoint, many things many things. You wonder in the morning whether the transportation would get there on time or not. Being on the overcrowded train and coming out messed up; wondering if someone stole your things (on the train) or not, all the rubbish on the streets and the filthy words we hear from men on the streets.....The stress at your work and the work hours and the salary, so many things. It's so expensive to rent an apartment with my salary. I won't be able to survive.[8]

Economic exclusion of youth is only one aspect of exclusion, and it is manifested in several factors as identified by Green et. al. (2016): 1) labour market exclusion, which makes it harder for some groups to join the labour market due to discrimination, the changing structure of the labour market, or not having the skills required to secure employment. 2) Poor quality jobs are another form of economic exclusion, where people get trapped in stagnant, low-end wages with unpredictable or excessive hours, lack of job security, lack of benefits and limited career

[7] Interviewee Raneen (Pseudonym), 25, Muslim, Female, Pharmacist Assistant, Tunis.

[8] Interviewee Sina (Pseudonym), 23, Muslim, Female, Dental Lab Technician, Tunis.

pathways. 3) Economic vulnerability, which leaves people unprotected when they go through financial emergencies, such as a job loss or a health crisis. 4) Economic isolation from opportunity, which occurs when there is segregation in neighborhoods, which leaves certain groups without access to jobs, quality schools and public services. 5) The inability to create jobs or to establish a business, as a result of lack of skills, lack of access to capital, and restrictions and poor infrastructure in certain areas.

Poverty and unemployment, which are plaguing Arab cities, have hit the youth harder. Arab youth, in particular, face the highest unemployment rates in the world at 22.9% (ILO, 2020), with women being hit the hardest at 42.5% (ibid.). Youth are less likely to find jobs compared to the adult population—the youth: adult unemployment ratio is 4.5 (ILO, 2017). Unemployment rates tell half the story, those who are employed are often faced with unstable hazardous working conditions, poor employment packages and poverty wages. Economic exclusion can both cause and reflect social isolation; residing in poorly resourced neighborhood can result in reduced economic opportunities, reduced social connections and reduced exposure to working role models and employers (Silver, 2007). Being condemned to economic exclusion can affect the realization of other rights. Marshall (1950) has discussed how rights are intertwined, he explains how minimum standards of living in terms of income, housing, health care and education are required to enable a person to defend their civil rights and to participate in the political arena.

> I dream about having a stable life and not needing to worry every month if I'm going to be able to cover my expenses or not. It's hard to be creative and to produce great things when you are worried all the time about the basics of living.[9]

Youth in Arab cities are not only suffering from economic exclusion, they are also excluded from political participation and from decision-making processes. Opportunities for Arab youth to engage in their communities, through shaping their space, influencing institutions that take decisions directly affecting their lives, political participation, whether in local or national elections, volunteering and voicing their needs are limited. For example, in Egypt pre the 2011 uprisings, only two of the then functioning twenty-six political parties had youth divisions (Sika, 2012),

[9] Interviewee Khalil (Pseudonym), 25, Muslim, Male, Dancer/Actor, Tunis.

and even after the revolution, political party participation among young Egyptians was only 0.3% (Sika, 2016).

Disillusioned by political parties and disappointed with their self-interests, corruption, domination by older generation and co-option by authoritarian regimes, young people prefer online activism, single-issue campaigns and youth-friendly spaces, rather than belonging to a political organization (Calder, et al., 2017). Almost one-third of Tunisian youth, for example, rate politicians as unintelligent and two-thirds rate politicians as dishonest (Arab Barometer VI cited in Yerkes, 2017). Lack of participation in formal politics means that youth remain excluded from decision-making processes and their needs remain invisible.

> I'm one of the people who don't feel positive about the elections. Honestly, there is nobody convincing you with their words or their actions. Look at the state of the country..... They (politicians) are fighting over positions. I don't feel there is hope, that's how I personally feel.[10]

Tunisians are frustrated with the level of corruption in their country and the dire economic situation that hasn't improved after the 2011 uprisings. They are disillusioned with the established political parties and their frustration was reflected in their voting tendencies in both the presidential and parliamentary elections.

> People even make jokes about the elections. They think that even if they vote nothing will change. I'm going to vote, but a lot of my friends are so disappointed by this country, they just want to leave.[11]

> Honestly, I didn't even follow the election campaigns. I feel it's an empty story. I heard a little of what was said, they talk about terrorism and security, and a little bit about the economy.[12]

In the presidential elections of 2019, the people of Tunisia expressed their disappointment with the established political class, by choosing two political independents, to move to the second round. Kais Saied won the elections, but that did not end the political instability in

[10] Interviewee Raneen (Pseudonym), 25, Muslim, Female, Pharmacist Assistant, Tunis.

[11] Interviewee Khalid (Pseudonym), 23, Muslim, Male, Unemployed, Tunis.

[12] Interviewee Raneen (Pseudonym), 25, Muslim, Female, Pharmacist Assistant, Tunis.

Tunisia, and the following parliamentary elections reflected a fragmented Tunisia.

The exclusion of youth is not only manifested on the political level, state policies towards youth have been consistently repressive. Youth are not only marginalized from political participation, policies in the four cities we are studying have been repressive towards youth to varying degrees. In the past four decades, the political economy of Arab cities has witnessed a marriage between a neoliberal economic model and the strengthening of the police authority in society to oppress dissent (Abdelrahman, 2017). The neoliberal restructuring process in Arab cities has necessitated greater investment in police forces, surveillance system and intimidation, in order to repress the protests of those harmed by its policies and in order to create a stable environment for investment and the accumulation of more wealth to those in power (ibid.).

> Youth they talk, but they cannot decide on anything and it takes more than protest to change the situation.... If I post anything on social media, if I criticize the government, I'm going to get arrested. Tribes are also beyond the law, and if I say this guy from this tribe is corrupt, I might get shot.[13]

After Sadat's ascendance to power, he fought youth activism and demonstrations in universities, by issuing repressive laws on the right to assembly, and the right to protest (Sika, 2016). However, the regime wanted to show that it respects its youth and wanted to hear their voices—but only the voices that conformed to the regime. Sika (2016) explains that there are two types of young people from the regime's perception, the first are the young people who obey the authoritarian structure, and can be co-opted within the regime, politically, economically and socially; on the other hand, are those who are perceived by the regime as radicals, such as the Revolutionary Socialists or the Muslim Brotherhood, whom the regime wanted to coerce by physical force and imprisonment, and to delegitimize them in the eyes of the public.

After the assassination of Sadat in 1981 by religious militants, a state of emergency was declared and the state security agency took charge at the expense of Egyptians civil liberties (Fahmy, 2012). People were detained on suspicion without a trial, some were released on paper while

[13] Interviewee Ayman (Pseudonym), 21,Muslim, Male, Architecture Student, Amman.

in reality they remained in custody, or were released only to be rearrested. The security agency interfered in appointments of heads of universities, heads of national newspapers, and public television stations to tighten the control on public discourse (ibid.). In the Mubarak era, youth with any hint of religious activism were crushed.

This trend of "controlling" the youth continued with the subsequent regimes. Even after the revolution, Morsi used the same policy of suppressing freedom of speech against young liberals who refused to cooperate with his regime. The rebellion that took place against Morsi later, was orchestrated by the Egyptian military figures supported by the United States, Israel, the UAE and Saudi Arabia, who opposed the Muslim Brotherhood being in charge in Egypt (Butler, 2020). The new Sisi government received support from Gulf donors, with a reform package proposed by the UAE government (ibid.). After the coup, Sisi followed the same neoliberal model adopted by his predecessors, while enjoying the support of the judiciary, the police, the Al Azhar religious authority, the media and the military (Joya, 2017). He proposed austerity measures and further privatization in all the sectors of the economy. This approach affects the poor the most, for example, while food and fuel prices increased by up to 78% in the middle of 2014, the government cut fuel and food subsidies (Abdelrahman, 2017). The iron fist with which the government is ruling the public means less chances of people protesting, or even questioning the government policies on media outlets.

The marriage between neoliberal policies and repressive authoritarian policies became very apparent in the Sisi era. Sika (2016) says that in the same manner as his predecessors, but with a more restrictive approach Sisi seeks to control the youth. The 2013 Protest Law mandates jail sentences for those who participate in any demonstration not approved by the interior ministry. Additionally the "Terrorist Entities law" employs loose terminology enabling the imprisonment of human rights activists, the "Social Networks Security Hazard Monitoring Operation" aims to utilize mass surveillance of digital activity. From Sadat's time onwards the regimes approach did not change neither towards the youth nor towards the economy (Sika, 2016). It was as Sika (2016: 7) labels it "old wine in new bottles." If anything, the 2011 uprising has allowed the ministry of interior to exploit the situation, and ask for an increase in its budget to quash any form of opposition and increase the power of the security bureaucracy (Abdelrahman, 2017). Sika (2016: 12) reports in an interview she did with a young political activist that "political participation by young activists today is totally banned. Today he is afraid to talk about politics or to talk about his political and security concerns with his friends or with citizens in cafés as he used to prior to and immediately after January 25, because he is afraid to go to jail".

Israel uses similar tactics to suppress Palestinian youth. The Israeli State policies reflect a strategy of control and containment of the Palestinian community, all the while pushing them out of the city's center. There is a heavy presence of Israeli forces around East Jerusalem, targeting Palestinian youth, subjecting them to intrusive body search, intimidation and imprisonment. In 2020 alone, 443 young Palestinian Jerusalemites were arrested in Jerusalem, 33% of them children under the age of 15 (ACRI, 2021). Palestinian Jerusalem youth feel prosecuted. "They stop me most of the times and you see the racism and the hatred in their eyes... they are going to search me anyway, whether I have anything or not."[14] Israel has installed two security points at the entrance to the Damascus gate of the Old City, constantly monitoring and provoking Palestinian youth as they enter the Old City of Jerusalem. "I feel that they enjoy searching me and humiliating me and they have all the power and the orders to shoot if they suspect someone, so all they need to have is just the suspicion and that's it."[15] Intimidation and harassment by the Israeli forces in Jerusalem ranges from "stop and frisk" to actual murder, as with the case of the autistic Palestinian young man Iyad Al Hallak, who did not respond to soldiers' orders to stop, and was shot dead as a result (B'Tselem, 2020).

> I'm one of the people who do not go near Damascus Gate, although I used to go to the Old City daily. My barber shop was there, but I don't go there anymore. The guy opened another shop outside the city, because he knows that young men are subject to harassment in that area. If I want to go down to the Old City, they (Israeli forces) will strip me, and touch my body, and (demand) put your hands up, put your hands down. I don't need to put myself in this situation, so I avoid the whole area.[16]

Although it is most severe in the case of East Jerusalem, harassment by police is also witnessed in other Arab cities. Police and municipal officials are part of the governing regime and their loyalty is with the regime and

[14] Interviewee Adam (Pseudonym), 21, Muslim, Male, Jerusalem.

[15] Interviewee Nadim (Pseudonym), 29, Christian, Male, Jerusalem.

[16] Interviewee Faris (Pseudonym), 27, Muslim, Male, Jerusalem.

not the city residents. It can be witnessed on the economic level in the rounding up of street vendors and pushing them out of the city center, or preventing and containing youth activism on the street, or random harassment of youth just for the mere reason of being present on the streets of the city, as several of the young interviewees expressed. "I get stopped by the police many times for no reason at all. Even when Im doing everything by the law. I get stopped and many times it's just harassment."[17] In some instances, not conforming to local cultural dress codes also provokes police intimidation. "I had long hair with bleached blond tips and I was stopped four five times by the police, and they would ask questions and they would imply that I am defaming my family name by looking like this."[18]

This feeling of insecurity and harassment by police was also reflected in the Power 2 Youth study of Calder, et al. (2017), who studied Arab youth in six countries. In their sample participants expressed the feeling that national police and security forces offer them little or no protection, and on the contrary, they exacerbate their physical insecurity. The police are serving the regime in power and not the people.

> A police officer once stopped the taxi I'm in and asked me to get down and then he let my taxi go. I gave him my documents and he said they are fake and I said check them on the system. He could have checked, while I was in the taxi, but he didn't. He humiliated me and let my taxi go. They just abuse their power. He was just bored and wanted to assert his power on me.[19]

Both harassment by the police and municipal policies limit the presence of youth in the city center. That is why when the Arab uprisings took place youth claimed the streets and sprayed the walls with graffiti to reclaim their space in the center. What we see today in those cities is the direct result of decades of policies that catered to the political and economic elite at the expense of the majority of the city population, with the youth population being particularly affected.

[17] Interviewee Samia (Pseudonym), 20, Muslim, Female, Art Student Amman.

[18] Interviewee Aziz (Pseudonym), 24, Muslim, Male, Visual Communications Student, Amman.

[19] Interviewee Ayman (Pseudonym), 21, Muslim, Male, Architecture Student, Amman.

Conclusion

This chapter started with a theoretical discussion on the importance of studying cities through the lenses of the "right to the city" and the "capabilities approach." The way the four cities we studied are governed has evidenced urban segregation and social exclusion of groups based on their social class and ethnic identities. Two issues contribute to the haphazard urbanization occurring in Arab cities. The first is the insufficiency of governmental policies that are based on actual local demands for housing, health care, infrastructure and economic opportunities (Stadnicki et al., 2014). The second is the policy of privatization, which has resulted in the spread of exclusive cities, industrial zones and touristic resorts that accumulate wealth for the few at the expense of the majority of city residents, who lack jobs, and access to quality education and various social services (ibid.). This emphasizes the importance of democratic local governance of cities and the adoption of pro-inclusive policies that are tailored and take into account the intricate specifics and needs of the different neighborhoods in cities. That is why they should not be a top-down policies or replicas of other cities' policies, but rather local policies that give space for grassroots participation and residents participation. Communities should be involved in the development plans, to make locally appropriate decisions about development.

Governance of cities is supposed to be independent of the central government, and led directly by elected officials with a mandate to develop their territory, who should be held accountable by their constituents. Examples of fair and democratic local elections show improved people's well-being in cities, in terms of the social services available, and also in terms of increasing revenues for cities. As long as people in the four cities we studied are not participating in the decision-making processes of how their cities are shaped, governed, and how resources are allocated, inequality and segregation will persist. Arab youth, in particular, are paying the price of these policies, with higher levels of unemployment and haphazard working and living conditions. They are not granted the opportunity to transition into adulthood in an environment that is safe, provides equal opportunities, supports representation in governance and provides them with the chance to have control over the space they are living in. Urban youth are facing exclusion on several fronts. The coming chapters will discuss these various forms of exclusion which include exclusion from high-quality education, exclusion from youth educational programs and activities, municipal services, functional affordable transportation lines, participation in the labour market and participation in the political arena and in shaping their cities. Youth also face harassment and intimidation by police and municipal officials, and are often excluded and pushed out of the city center. This is manifested in laws preventing freedom of speech and freedom of assembly, and in the way youth are included in the policy-making processes only when they conform with the regime narrative.

Bibliography

Abed, A. Tomah, A., & Dumour, D. (2015) Assessment of slums' upgrading interventions: Case study Jabal Al-Natheef, Amman, Jordan. *Innovative Systems Design and Engineering, 6*(6), 1–8.

Abdelrahman, M. (2017). Policing neoliberalism in Egypt: The continuing rise of the 'securocratic' state. *Third World Quarterly, 38*(1), 185–202.

Adely, F. (2009). Educating women for development: The Arab human development report 2005 and the problem with women's choices international. *Journal of Middle East Studies, 41*(1), 105–122.

AFP. (2021). *Egypt's Sisi looks to new desert capital to cement legacy.* https://www.rfi.fr/en/middle-east/20211114-egypt-s-sisi-looks-to-new-desert-capital-to-cement-legacy. Accessed 10 March 2022.

Anand, P., Hunter, G., & Smith, R. (2005). Capabilities and well-being: Evidence based on the sen-nussbaum approach to welfare. *Social Indicators Research, 74*(1), 9–55.

Al-Hamarneh, A. (2019). "Right to the city" in the Arab world: Case studies from amman and Tunis metropolitan areas. In A. Al-Hamarneh, J. Margraff & N. Scharfenort (Eds.), *Neoliberale urbanisierung: Stadtentwicklungsprozesse in der arabischen Welt* (pp. 185–212). Transcript Verlag.

Al-Khalidi, S. (2017). *Jordan holds local elections in step to devolve powers.* https://www.reuters.com/article/us-jordan-election-idUSKCN1AV2EK. Accessed May 1 2022.

Asali Nuseibeh, R. (2015). *Political and social exclusion in Jerusalem: The provision of education and social services.* Routledge.

Asali Nuseibeh, R. (2019). Palestinian women teachers in East Jerusalem: Layers of discrimination in the labor market. *The Middle East Journal, 73*(2), 207–223.

Association for Civil Rights in Israel (ACRI). (2021). *East Jerusalem facts and figures.* https://www.english.acri.org.il/post/__283. Accessed 1 May 2022.

Bayat, A., & Biekart, K. (2009). Cities of extremes. *Development and Change, 40*(5), 815–825.

B'Tselem. (2017). *East Jerusalem*. https://www.btselem.org/jerusalem. Accessed 1 May 2021.

Ben-Hillel, Y. (2013). *The legal status of East Jerusalem*. The Norwegian Refugee Council. https://esa.un.org/unpd/wup/Download/. Accessed 1 May 2022.

Bellamine, Y. (2015). A quand un projet de loi de décentralisation? *Nawaat*. https://nawaat.org/portail/2015/01/22/a-quand-un-projet-de-loi-de-decentralisation-1e-partie/.

Betselem. (2020). *Palestinians killed by Israeli security forces in the West Bank, since Operation Cast Lead*. https://www.btselem.org/statistics/fatalities/after-cast-lead/by-date-of-death/westbank/palestinians-killed-by-israeli-security-forces. Accessed 15 March 2021.

Blake, J. Bartels E., Efron S., & Reiter Y. (2018). *What might happen if Palestinians start voting in Jerusalem municipal elections? Gaming the end of the electoral boycott and the future of city politics*. RAND Corporation.

Bourkinas, I. (2021). *Income inequality convergence across Egyptian Governorates*. Economic Research Forum. https://theforum.erf.org.eg/2021/03/08/income-inequality-convergence-across-egyptian-governorates/. Accessed 20 May 2022.

Brooks-Gunn, J., Duncan, G. J., Klebanov, P. K., & Sealand, N. (1993). Do neighborhoods influence child and adolescent development? *American Journal of Sociology, 99*(2), 353–395.

Butler, D. (2020). *Egypt and the Gulf*. https://www.chathamhouse.org/2020/04/egypt-and-gulf. Accessed 16 May 2022.

Calder, M., MacDonald, R., Mikhael, D., Murphy, E., & Phoenix, J. (2017). Marginalization, young people in the south and east Mediterranean, and policy: An analysis of young people's experiences of marginalization across six SEM countries, and guidelines for policy-makers (Power2Youth Working Paper No.35).

CAPMAS. (2021). Statistical yearbook. In *Central Agency for Public Mobilization and Statistics*.

Carniege. (2018). *Tunisia's 2018 municipal elections*. https://carnegieendowment.org/files/2018TunisiaElectionsInfographic.pdf. Accessed 1 May 2021.

Chaaban, J. (2009). Youth and development in the Arab countries: The need for a different approach. *Middle Eastern Studies, 45*(1), 33–55.

Chabbi, M. (2010). Tunis: Why housing is inadequate. *A planet for life*. http://regardssurlaterre.com/en/tunis-why-housing-inadequate. Accessed 1 May 2022.

Côté, J. (2014). Towards a new political economy of youth. *Journal of Youth Studies, 17*(4), 527–543.

Council of Europe. (2018). The first municipal elections in Tunisia, A starting point for further decentralization. *Congress of Local and Regional Authorities*.

https://www.coe.int/en/web/congress/-/successful-municipal-elections-as-a-starting-point-for-further-decentralisation-in-tunisia. Accessed 20 May 2022.

Daher, R. (2013). Amman's rising "kitsch syndromes" and its creeping vernacularized urban landscapes. *Ethnologies, 2*(35), 55–75.

Deneulin, S. (2014). Creating more just cities: The right to the city and capability approach combined. Bath Papers in International Development and Well-being, No. 32, University of Bath, Centre for Development Studies (CDS), Bath.

Department of Statistics. (2021). *Population census*. http://dosweb.dos.gov.jo/ar/population/population-2/. Accessed 20 March 2022.

Dumper, M. (1997). *The politics of Jerusalem since 1967*. Columbia University Press.

Egdell, V., & McQuaid, R. (2014). Supporting disadvantaged young people into work: Insights from the capability approach. *Social Policy & Administration, 50*(1), 1–18.

Fahmy, H. (2012). An initial perspective on "The winter of discontent": The root causes of the Egyptian revolution. *Social Research, 79*(2), 349–376.

France, A., & Threadgold, S. (2016). Youth and political economy: Towards a bourdieusian approach. *Journal of Youth Studies, 19*(5), 612–628.

Ginwright, S., & James, T. (2002). From assets to agents of change: Social justice, organizing, and youth development. *New Directions for Youth Development, 2002*(96), 27–46.

Glaeser, E. L., Resseger, M. G., & Tobia, K. (2009). Inequality in cities. *Journal of Regional Science, 49*(4), 617–646.

Greene, S., Pendall, R., Scott, M., & Lei, S. (2016). *Open cities: From economic exclusion to urban inclusion*. Urban Institute.

Haaretz. (2022). Family unification bill meant to stop palestinian 'creeping right of return,' Israel's Shaked Says. https://www.haaretz.com/israel-news/shaked-family-unification-bill-meant-to-stop-palestinian-creeping-right-of-return-1.10601224. Accessed 1 May 2022.

Halabi, Y. (2014). Democracy, clan politics and weak governance. *Israel Studies, 19*(1), 98–100.

Hanssen, J. (2018) Municipal Jerusalem in the age of urban democracy: On the difference between what happened and what is said to have happened. In A. Dalachanis & V. Lemire (Eds.), *Ordinary Jerusalem, 1840–1940: Opening archives, revisiting a global city*. Brill.

Harris, M. (2015). *Jordan's youth after the Arab Spring*. https://www.jstor.org/stable/pdf/resrep10167.pdf?refreqid=excelsior%3A930410c20dee9994babaee809721cb01&ab_segments=&origin=. Accessed 1 May 2021.

ICAHD. (2022). *The Israeli committee against house demolitions*. Homepage. https://icahd.org/. Accessed 5 March 2022.

ILO (International Labour Organization). (2020). *Global employment trends for youth 2020: Arab states*. https://www.ilo.org/wcmsp5/groups/public/---dgreports/---dcomm/documents/briefingnote/wcms_737672.pdf. Accessed 16 May 2022.

International Labour Organization. (2017, November). *Global Employment Trends for Youth 2017*.

Ir Amim. (2021). *Planned negligence: How Palestinian neighborhoods disappeared from Jerusalem's current & future urban planning policies*. https://www.ir-amim.org.il/sites/default/files/Planned%20Negligence_June2021.pdf. Accessed 25 Februray 2022.

Israel Ministry of Interior. (2018). *Results of municipal elections*. https://sevev.moin.gov.il. Accessed 15 June 2021.

Jabareen, Y. (2014). 'The right to the city' revisited: Assessing urban rights. The case of Arab Cities in Israel. *Habitat International, 41*, 135–141.

Jonathan, B., Bartels, E., Efron, S., & Reiter, Y. (2018). *What might happen if Palestinians start voting in Jerusalem municipal elections? Gaming the end of the electoral Boycott and the future of city politics*. RAND Corporation.

Joya, A. (2017). Neoliberalism, the state and economic policy outcomes in the post-arab uprisings: The case of egypt. *Mediterranean Politics, 22*(3), 339–361.

Khalifa, M. (2011). Redefining slums In Egypt: Unplanned versus unsafe areas. *Habitat International, 35*(1), 40–49. https://doi.org/10.1016/j.habitatint.2010.03.004

Lamont, M., & Molnár, V. (2002). The study of boundaries in the social sciences. *Annual Review of Sociology, 28*(1), 167–195.

Lefebvre, H. (1996 [1968]). The right to the city. In H. Lefebvre (Ed.), *Writings on cities* (E. Kofman & E. Lebas, Trans.) (pp. 63–184). Blackwell.

Loewert, P., & Steiner C. (2019). The new administrative Capital in Egypt: The political economy of the production of urban spaces in Cairo. *Middle East-Topics & Arguments, 12*, 66–75.

Lopez-Fogues, A., & Cin, F. M. (2017). *Youth, gender and the capabilities approach to development: Rethinking opportunities and agency from a human development perspective*. London, Routledge.

Mandour, M. (2021). *The sinister side of Sisi's urban development*. Carnegie endowment for international peace. https://carnegieendowment.org/sada/84504. Accessed 15 February 2022.

Marcuse, P. (2012). Whose right to what city? In N. Brenner, P. Marcuse, & M. Mayer (Eds.), *Cities for people, not for profit: Critical urban theory and the right to the city* (pp. 24–41). Routledge.

Marshall, T. H. (1950). *Citizenship and social class* (T. Bottomore, Ed.). Pluto.

Mayer, M. (2012). The "right to the city" in urban social movements. In N. Brenner, P. Marcuse, & M. Mayer (Eds.), *Cities for people, not for profit: Critical urban theory and the right to the city* (pp. 63–85). Routledge.

Monbiot, G. (2016). *Neoliberalism—The ideology at the root of all our problems*. https://www.theguardian.com/books/2016/apr/15/neoliberalism-ideology-problem-george-monbiot. Accessed 1 May 2021.

Morange, M., & Spire, M. (2019). The right to the city in the global south. perspectives from Africa. *Cybergeo*.

Muasher, M. (2011). *A decade of struggling reform efforts in Jordan: The resilience of the rentier system*. Carnegie Endowment.

Nicolenyi, C. (2020). The 2018 municipal elections in Jerusalem: A take of fragmentation and polarization. *Contemporary Review of the Middle East, 7*(1), 6–24.

Purcell, M. (2002). Excavating Lefebvre: The right to the city and its urban politics of the inhabitant. *GeoJournal, 58*(2–3), 99–108.

Purcell, M. (2013). Possible worlds: Henri Lefebvre and the right to the city. *Journal of Urban Affairs, 36*, 141–154.

Ranko, A., Lotz, M., Ghazaleh, H. A., & Imke, H. (2015). The municipal and governorate council elections of August 2017: Decentralization efforts in Jordan. *Konrad-Adenauer-Stiftung*. https://www.kas.de/c/document_library/get_file?uuid=7ef4a752-5c7e-7627-6600-4b99bd35fbf1&groupId=252038. Accessed 1 May 2021.

Schechla, J. (2014). The right to the city: Jerusalem. *Habitat international coalition—Housing and land rights network*. https://www.hlrn.org/img/publications/jerusalem%20profile_10-2014-final.pdf. Accessed 1 May 2021.

Schechla, J. (2015). The right to the city: Cairo. In *The land and its people: Civil so- ciety voices address the crisis over natural resources in the Middle East/North Africa* (pp. 133–144). Housing and Land Rights Network.

Séjourné, M. (2009). The history of informal settlements. In R. Kipper & M. Fischer (Eds.), *Cairo's informal areas: Between urban challenges and hidden potentials* (pp. 17–21). Norprint SA.

Sen, A. (1985). Well-being, agency and freedom: The dewey lectures 1984. *The Journal of Philosophy, 82*(4), 169.

Sen, A. (1992). *Inequality re-examined*. Clarendon Press.

Sen, A. (1999). *Development as freedom*. Knopf.

Sika, N. (2012, August). Youth political engagement in Egypt: From abstention to uprising. *British Journal of Middle Eastern Studies, 39*(2), 181–199.

Sika, N. (2016). *Youth civic and political engagement in Egypt* (Working paper no. 18). ISSN 2283–5792.

Sims, D. (2014). *Egypt's desert dreams: Development or disaster?* (1st ed.). The American University in Cairo Press.

Shlomo, O. (2016). Between discrimination and stabilization: The exceptional governmentalities of East Jerusalem. *City, 20*(3), 428–440.

Shlomo, O. (2017). The governmentalities of infrastructure and services amid urban conflict: East Jerusalem in the post Oslo era. *Political Geography, 61*, 224–236.

Shmid, C. (2012). Henri Lefebvre, the right to the city, and the new metropolitan mainstream. In N. Brenner, P. Marcuse, & M. Mayer (Eds.), *Cities for people, not for profit: Critical urban theory and the right to the city* (pp. 42–62). Routledge.

Silver, H. (2007). Social exclusion: Comparative analysis of Europe and Middle East Youth. *SSRN Electronic Journal.*

Stadnicki, R., Vignal, L., & Barthel, P. A. (2014). Assessing urban development after the 'Arab Spring': Illusions, evidence of change. *Built Environment, 40*(1), 5–13.

Stambouli, F. (1996). Tunis city in transition. *Environment and Urbanization, 8*(1), 51–63.

Tadamun. (2015). *Egypt's new cities: Neither just nor efficient*. http://www.tadamun.co/egypts-new-cities-neither-just-efficient/?lang=en#.Y1I8xS2l01I. Accessed 26 December 2021.

United Nations, Department of Economic and Social Affairs, Population Division. (2019). *World urbanization prospects: The 2018 revision* (ST/ESA/SER.A/420). United Nations.

UN Habitat. (2011). *Cairo: A city in transition* (Cities and Citizens Series, Bridging the Urban Divide). Cairo: The American University in Cairo, UN Human Settlements Programme, UN Habitat for a Better Urban Future.

UN Habitat. (2017). *The right to the city and cities for all*. https://uploads.habitat3.org/hb3/Habitat%20III%20Policy%20Paper%201.pdf. Accessed 1 May 2021.

UN Habitat and the League of Arab States. (2016). *Towards an Arab urban agenda*. https://unhabitat.org/sites/default/files/documents/2019-05/towards_an_arab_urban_agenda.pdf. Accessed 22 March 2022.

United Cities and Local Governments. (2012). https://www.uclg-cisdp.org/sites/default/files/CISDP%20Carta-Agenda_ENG_0.pdf. Accessed 22 March 2022.

United Nations. (2018). World urbanization prospects: The 2018 revision. Department of Economic and Social Affairs, Population Division (2018). https://esa.un.org/unpd/wup/Download/. Accessed 1 May 2022.

UNFPA. (2020). *Youth participation and leadership*. https://arabstates.unfpa.org/en/topics/youth-participation-leadership. Accessed 22 March 2022.

Van Ham, M., Hedman, L., Manley, D., Coulter, R., & Östh, J. (2013). Intergenerational transmission of neighbourhood poverty: An analysis of neighbourhood histories of individuals. *Transactions of the Institute of British Geographers, 39*(3), 402–417.

Van Ham, M., Tammaru, T., & Janssen, H. (2018). A multi-level model of vicious circles of socio-economic segregation. *OECD Divided Cities* (pp. 135–215). OECD Publishing.

World Bank. (2017). *Public expenditure and financial accountability (PEFA) assessment*. Greater Amman Municipality, The Hashemite Kingdom of Jordan. https://webcache.googleusercontent.com/search?q=cache:YJNwt1Dc5sgJ:https://documents1.worldbank.org/curated/fr/292151513880097520/12§4-revised-GAM-PEFA-Report-FINAL-Nov-10.docx+&cd=1&hl=en&ct=clnk&gl=et. Accessed 1 May 2021.

Yerkes, S. (2017). *Where have all the revolutionaries gone?* Brookings. https://www.brookings.edu/wp-content/uploads/2017/03/cmep_20160317_where_have_revolutionaries_gone.pdf. Accessed 15 May 2022.

Yerkes, S., & Muasher, M. (2018). *Decentralization in Tunisia: Empowering towns, engaging people*. Carnegie Endowment for International Peace. https://carnegieendowment.org/2018/05/17/decentralization-in-tunisia-empowering-towns-engaging-people-pub-76376. Accessed 1 May 2021.

CHAPTER 3

The Political Economy of Youth Exclusion

Introduction

Middle Eastern studies that analyse authoritarian regimes in the Arab world and attempt disassociate politics from economics, or that use Orientalist approaches that link Arab culture and Islam to authoritarianism, distract from the real issues perpetuating oppression (Zemni, 2017). These Orientalist stereotypes have distracted researchers from looking into how the global capitalist system fuels and perpetuates inequalities and oppression in the region.

The labour markets in Tunisia, Egypt and Jordan, all exhibit similar trends. They have large but shrinking public sectors, that offer desirable stable employment conditions. Public sector jobs offer greater job security, higher wages and higher benefits and are magnets for the more qualified labour force (IMF, 2012). In Jordan the government absorbs a large amount of the workforce. Working for the government is attractive to Jordanians, as it offers stable employment, reasonable salary, benefits and social security, especially since only 42% of the labour force in Jordan is covered by the social security system (Taghdisi-Rad, 2012). However, government and military jobs are usually reserved for Jordanians from the East Bank (i.e. excluding Palestinians). Jordanians who are originally Palestinians from the West Bank, who make up over half of the population are more concentrated in private sector employment. While privatization

R. A. Nuseibeh, *Urban Youth Unemployment, Marginalization and Politics in MENA*, Middle East Today,
https://doi.org/10.1007/978-3-031-15301-3_3

eroded the public sector, which is the desired job destination of East Bankers, it heightened the sectarian divide between East and West Bankers (Pelham, 2011). Similar trends can be seen in Tunisia and Egypt; government jobs are mostly available for politically significant groups. Assaad (2014) explains that public sector jobs have been used by authoritarian regimes as a bargaining tool with politically significant groups, to provide them with well-compensated jobs in the bureaucracy and the security forces in exchange for political loyalty. This has meant that the formal private sector remains weak. What is more, with the shift towards liberalization policies, this sector is also rife with corruption, as the business elite and privileged groups connected to the ruling parties acquire opportunities to create monopolies. These economies are also plagued with restrictive employment contract laws, sluggish improvements in labour laws, with more advantages and an upper hand to big private companies and corporations at the expense of mass labourers (Al Azzawi & Hlasny, 2020). These economies also have dominant service sectors, which have lower labour absorption potential (Dimova & Stephan, 2020). This has led to a growing informal private economy, that offers precarious working conditions and irregular wages, with an over-representation of youth from lower wealth quintile, especially in Egypt and Tunisia, where the children of less wealthy and educated parents start out in precarious jobs and are unlikely to ever attain formal employment (Al Azzawi & Hlasny, 2020).

In the case of East Jerusalem, the double effect of neoliberalism and urban colonialism results in massive restrictions on Palestinians in the economic sector. Clarno (2017: 323) has defined this as "neoliberal colonization," where the intersection of both neoliberal economic policies, and colonial expansions into Palestinian spaces have resulted in devastating effects on the Palestinian community. In the context of East Jerusalem, the effects are visible in the high levels of poverty (ACRI, 2021), limited development and expansion of Palestinian businesses, limited job opportunities for highly educated Palestinians, low average salaries and the clustering of Palestinians in low-skilled professions.[1]

[1] "Jewish men and Arab men work in very different economic sectors: among Jewish men the main sectors were education (16%), local and public administration (13%), trade (9%), professional and scientific services (9%), and human health and social work services (8%), while among Arab men the four salient sectors were construction (19%), trade (19%), transportation and storage services (12%), and accommodation and food services (11%)" (Korach & Choshen, 2021: 74).

Bonvin and Galster (2010) argue that looking at the labour market through the lens of the "capability approach" means that the objectives of social policy and public actions should be directed towards developing people's skills and enhancing their freedoms to choose the work that they have reason to value, meaning capability for work. They say that we should look at two elements; "opportunity freedom" and "process freedom" "Opportunity freedom" in the labour market has two prongs, first the supply side which entails providing people with the training and the skills needed to be able to access the labour markets, while at the same time eliminating the discriminatory processes that affect people's ability to access the market, such as gender, race and nationality issues (Bonvin & Galster, 2010: 72). The second prong is the demand side, which means the opportunities that are available in the market for people; opportunities which are created by the collective effort of the state apparatuses, market actors and civil society institutions. Bonvin and Galster (2010) explain that this does not mean that people should be passive actors waiting for training and then choosing from the list of employment options available to them, instead they should be active in the decision-making processes, and this is what they mean by "processes freedom", where people have a say in the strategies and the implementation of those strategies in the labour market. Rather than a top-down approach to public policy, there needs to be a reflexive, participatory approach, where people can participate in decision-making processes related to labour market opportunities. Most of the literature on Arab youth unemployment discusses the human capital approach and how these youth are not getting the "needed training" and skills. They are treated as passive receivers, with no say or agency in shaping the labour markets or having the right to shape their space in ways that they deem valuable. Bonvin and Farvaque (2007) have argued that going beyond the human capital narrative of equipping people for the labour market means that there is a need to create a capability-friendly institutional environment, which works on enhancing human capital, social connections and people's capabilities in the labour market, so they can participate in the shaping of their markets.

The "capability approach" has been able to challenge the economic-centric human capital approach to education. This revaluation process is increasingly happening in the labour markets, with more scholars advocating for the use of the "capability approach" to frame participation in the labour markets as a capability, where people can have more choice (Bussi & Dahmen, 2012). However, taking a look at the economies in our four contexts, youth do not have much choice; especially given the wide gaps in opportunities according to gender, age, wealth and ethnicity. Although this youth cohort is more educated, have higher expectations, and is more exposed to the world than previous generations, what is available for youth in the local market in terms of quantity or quality of job opportunities is limited and the capabilities they have do not enable them to compete in the global market, which means they will fail to become independent and remain subordinated to the kinship structures where elders and males hold the power, especially over females (Calder et al., 2017).

Neoliberal Policies and Oppressive Regimes

Urban marginalization encompasses various conditions, such as living in informal settlements, lack of access to social services or transportation, precarious working conditions and lack of job and educational opportunities. It also includes being subject to discriminatory policies such as being forcibly displaced and pushed away from the center and from spaces where there is opportunity, as well as being disempowered and not having an input in the design and the functioning of one's urban space.

Most Arab cities have not developed their infrastructure, transportation lines, education, health and work opportunities in ways that matches their population growth. Public transportation in particular has a major effect on the urban economy of cities. For example, although Cairo and Beijing share the same population size, the GDP of Cairo for the year 2015 was a quarter of that of Beijing (Elgendy & Abaza, 2020). Cairo with its large population has a limited urban metro network of 3 lines on 78 km and a bus network of 11 lines with 105 buses (ibid), in comparison, Bejing has a subway with 25 lines, and its has over 1200 bus routes (Nguyen, 2022). This limited investment in the city transportation network in Cairo not only contributes to air pollution and traffic congestion, but also affects the productivity and the economic development of the city. It would surely make a difference to the accessibility of the marginalized to the city-center if there is an efficient and free (or cheap) public transport system to bring people into the city center. Amman has also failed to sufficiently invest in its public transportation, leading to higher numbers of private cars, pollution and congestion, and harm to the urban economy (Elgendy & Abaza, 2020).

Bogaert (2013) explains that just like capitalist companies, urban territories are being treated in the same competitive logic to maximize economic growth and capital accumulation, which entails the prioritization of tourist projects, foreign investments in real estate projects, luxury resorts and industrial zones, at the expense of the majority of city-dwellers, and their economic and social well-being. The ideology of "how can we sell the city" trumps the ideology of "how can we make the best possible city for its residents" (Bogaert, 2013: 227). These projects often hike prices and push middle-class and poor segments of society out of the city center and into the periphery, or they integrate them into these projects as cheap laborers, whose work conditions keep them under the poverty-line. Sika (2018) points out that although private investment increased enormously in Egypt from the late 1980's onwards, the economic sectors towards which these investments were directed generated few jobs, and most investments reinforced the dependence on oil, tourism and real estate developments, which do not add enough to employment opportunities. As for the private sector, which according to the World Bank, IMF and UNDP policies, was supposed to absorb new entrants into the labour market, it did not, instead leading to two-thirds of young people being employed in the informal sector, depriving them of employment benefits, social security and stability (Sika, 2018).

> There should be more job opportunities for university graduates; there aren't any. Today people are saying "why should I go to university, when I will anyway not find employment after graduation?.[2]

The governance model in Egypt has been largely shifted towards privatization, which started in the Sadat era in the 1970s. Sadat's adoption of open market policies, privatization and deregulation was meant to disassociate his regime from his predecessor Jamal Abdel Nasser's "socialist legacy" (Fahmy, 2012). This has meant a shift away from state-planned economies and regulated public sector enterprises, such as agriculture and branches of industry to a private sector-led model of development (Joya, 2017). The idea of this reform was to open the doors to private investment to get rid of the bloated wasteful government sector. However, this resulted instead in a corrupt capitalist system, where people who were politically connected benefited from the privatization of public assets and

[2] Interviewee Haya (Pseudonym), 19, Christian, Female, Economics Student, Cairo.

amassed fortunes at the expense of workers who experienced wage stagnation and rising unemployment (Joya, 2017). Mubarak continued with Sadat's policies, and in the nineties, Egypt witnessed a decline in public sector investments, further privatization of state-owned enterprises, and the relinquishing of the "government employment guarantee scheme" (Barsoum, 2015). For example, a law that came into effect in 1997 revoked the previously state-determined rents of small farmers, leading to an increase of rents by up to 400% along with the dispossession of millions of peasant families, and thereby to their migration to urban centers, in particular Cairo (Abdelrahman, 2017; Hanieh, 2013).

While these reforms did restructure the bureaucratic state, privatization remained state-led, with the army controlling 40% of the economy and with a significant number of officers sitting on boards of directors of a wide range of state-owned public utilities; infrastructure, gas and oil industries, shipping companies, real estate and housing companies, and vast agricultural farms (Joya, 2017; Sayigh, 2012). In the name of "national security," the military has used its privileged position since 1952 to protect its monopolistic practices and its budgets from any form of external scrutiny (Abdelrahman, 2017). This overlap between economic and political power has prevented the growth of an independent competitive private sector that is able to create jobs for the youth (Malik & Awadallah, 2013). As for jobs that were available on the market, they were informal with low wages and bad conditions, which debunked the rationale that free market and privatization would lead to the creation of jobs and lead to more growth.

Although in 2009 the World Bank evaluation report stated that between 1999 and 2007 Egypt's economic performance improved substantially (IEG, 2009), this growth was mostly derived from labour remittances, Suez Canal fees and high energy prices. Only a small number of the population were benefiting from this growth, while 44% of Egypt's population were suffering under the poverty line (Abdelrahman, 2017).

Since 1986, Jordan has also pursued market-oriented policies, following the global trends of laissez-faire policies. It has permitted the privatization of public services, reduced barriers to foreign investment, created tax-free zones, and formulated trade agreements with the USA, Europe and throughout the regio. It has joined the Greater Arab Free Trade Area (GAFTA) in 1998, and formulated the Agadir agreement with Tunisia, Morocco and Egypt in 2004. Additionally entered into an association agreement with the European Union 1997, and with the EFTA (European Free Trade Association) countries in 2002 (Royal Science Society, 2013). It also entered into a Qualified Industrial Zones Agreement (QIZ) with the USA in 1996, whereby goods manufactured in the zones receive quota-free and duty-free access to the US market. But the agreement came with strings attached (International Trade Administration, undated); if a company seeks to export its products to the US under this agreement, the product needs to be manufactured in the industrial zone, 35% of the product should have Jordanian content, of which 11.7% must come from the industrial zone, 8% from Israel (7% for high tech goods) and the remainder may be fulfilled by content from a Jordan QIZ, Israel, the USA or the West Bank or Gaza, and this process should be regulated by a joint committee consisting of Jordanians, Israelis and an observer from the US (Taghdisi-Rad, 2012). Despite these stringent conditions, the QIZ has been to Jordan's advantage as figures show that through the QIZ Agreement Jordan was able to export $1.14 billion of goods in 2007 and it has employed over 46,000 workers (ibid.). However, 88% of capital invested is from foreign firms, and in their drive to reduce the costs of production, they have depressed wages and downgraded working conditions, resulting in over half of the workers in the zones being migrant workers in desperate need to find employment (ibid.). The government's strategy of attracting foreign investment and generating growth to create jobs was therefore not sufficient in reducing unemployment. Looking at the macro-level, the policies were successful in attracting foreign capital, especially in the real estate sector, and in increasing economic growth, which was reflected in the steady increase of its GDP between 2000 and 2010 (Brown et al., 2014). However, this economic growth was not accompanied by increasing labour force participation. Taghdisi-Rad (2012) argues that the liberalization of the economy has actually resulted in more job destruction than job creation. The economic liberalization and export promotion processes in Jordan have particularly focused on low manufacturing sectors, creating unskilled jobs in poor working conditions, jobs that are mostly attractive to migrant workers, especially those coming from Egypt, who are willing to accept lower wages and less attractive work packages (Brown et al., 2014). Employers also prefer to hire migrant workers, since they are on temporary contracts, and employers are not obliged to contribute to social security or severance pay for these workers. In 2014, Jordan hosted between 300,000 and 400,000 non-refugee foreigners to work primarily as unskilled labor force in construction and domestic work (ibid).

Tunisia's Ben Ali regime adopted the same neoliberal economic approach. It privatized national Tunisian companies, such as telecommunications, transport and cement industries (Aleya-Sghaier, 2012). The president and his cronies enriched themselves by using their political leverage, using this privatization process to accumulate wealth and channel most of it outside the country (Murphy, 2017). The high scale of corruption in Ben Ali's regime destroyed market competition, hindering the emergence of an autonomous and competitive market, and caused reluctance among investors to venture into the Tunisian market (Paciello, 2011). Ben Ali's political and economic model reflected the same marriage between a neoliberal economic model and the coercive oppressive measures to quell revolt. The regime "staged democratic performances" largely driven by its desire for Western investments and alliances (Murphy, 2017). It legalised select opposition parties, which assumed the role of "loyal opposition", while banning genuine opposition from the communist left or the Islamist right (Murphy, 2017: 682). The true face of the regime was reflected in the surveillance, harassment, imprisonment and torture of human rights activists, journalists and members of the true opposition (Kausch, 2009). The economic liberalizing policies in Tunisia were accompanied with social programs, which aimed to alleviate poverty and unemployment (Zemni, 2017). However, Paciello (2011) argues that these expenditures were used as a tool of control by the government, who determined which families to exclude from the state's social benefits, based on political loyalty and silence. Also, the majority of economic investments in human capital and infrastructure were concentrated in the northern cities, but even in the northern affluent cities there were and still are pockets of deprivation. After the ousting of Ben Ali's regime, the new government started its tenure with many challenges, inheriting a weak economy, which was then hit by a pandemic. Tunisia's GDP fell by 8.6% in 2020 compared to the previous year, and its public debt had reached 87.6% of the GDP at the end of 2020 (Dridi, 2021). Its external debt has now reached 30 billion dollars (ibid.).

The Israeli experience with neoliberalism has included similar privatization policies as seen elsewhere, which included free trade, privatization, deregulation, cuts on social spending, attacks on unions and welfare programs. As with the other examples, this resulted in exacerbating inequality in the country and put Israel high on the measure of inequality after the United States (Zureik, 2020). Added to the common privatization policies, that impact the lives of both the Palestinian and Israeli citizens of the State of Israel, Israeli neoliberalism has had dire effects on Palestinians living in the Occupied Palestinian Territories. This has been reflected in the expansion of settlements built illegally on Palestinian land, ethnic segmentation of the West Bank, and the adoption of privatized security campaigns against the Palestinian population, that breach their human rights and dignity (ibid.). Policing, checkpoints, issuing of travel permits have been either partially or fully privatized, to save the government from the cost of the occupation, and also absolve the military from breaches of human rights as they impose daily humiliations on Palestinians in the Occupied Territories. The private security sector is one of the fastest-growing industries in Israel, accounting for $200 million a year (ibid). Half of the checkpoints between the West Bank and Gaza have been privatized, as well as six prisons holding political Palestinian prisoners, which are all run for profit with complete disregard to human rights and dignity (ibid). Even the way that the Palestinian Authority functions today is a reflection of the Israeli authorities outsourcing and privatizing the management of Palestinian lives in the Palestinian Occupied Territories (Gordon, 2008, Hever 2008 cited in Zureik, 2020). Former Prime Minister Itzhak Rabin has said that the most effective way to ensure Israel's security, and to limit Palestinian resistance is to encourage the establishment of an undemocratic Palestinian regime, that would disregard human rights, democratic procedures and the rule of law.

> I prefer the Palestinians to cope with the problem of enforcing order in Gaza. The Palestinians will be better at it than we, because they will allow no appeals to the Supreme Court, and will prevent the Association for Civil Rights [in Israel] from criticizing conditions there, by denying it access to the area. They will rule there by their own methods, freeing - and this is most important -IDF soldiers from having to do what they will do (Yedi'ot Aharonot, September 7, 1993 cited in Beinin, 1998: 30)

Although in its inception stage the Palestinian Authority gave people the promise of self-determination, once in power, it started allowing neocolonial relations of production with Israel to bolster its own power and secure privileges for the national elites, rather than achieve a political and just solution for the Palestinian people (Khalidi & Samour, 2011). The Palestinian Authority program is built on a neoliberal model, where it basically offers an economic program without any strategy for ending the illegal Israeli Occupation. This is also pushed by the US agenda, that is primarily focused on securing a stable environment in the Middle East for its investments and supply of oil (Khalidi & Samour, 2011). It was also reflected in the Abraham Accords[3] that sought to integrate Israel in the region, in exchange for economic prosperity for the Palestinians, while ignoring the Occupation. Although Palestinian officials have denounced Israel's "economic peace", the reality on the ground is that the Palestinian Authority relies on Israel for any economic agenda. They don't have an independent central bank, they have no means to reduce interest rates and inflation or to set a competitive currency exchange rate in support of export-led economic growth, they cannot independently reduce tariff rates or value added taxes, they do not even have the power to stop Israel's illegal settlement expansion, or confiscation of Palestinian land and resources (Khalidi & Samour, 2011). The Oslo Accords[4] did not reduce the power asymmetries between the Palestinians and Israelis, in fact the Paris Economic Protocols, which are the economic annex of the Oslo Accords institutionalized Palestinian economic dependency on Israel, by granting Israel exclusive power over key Palestinian Authority economic resources, such as monetary policies, trade and fiscal revenues, labour flows and industrial zoning rights (Dana, 2020).

Even if the Palestinian Authority had opted not to follow neoliberal economic policies, pressures from donor organizations, the US and Israel (in control of their economy, as of every other aspect of Palestinian lives) would have set it back in this direction. The situation in the Occupied Palestinian Territory as defined by Clarno (2017) is a form of neoliberal colonization, where both aggressive settler colonization and neoliberal capitalism are severely impacting the realities of Palestinian life. Political and profit-making activities in the Palestinian Occupied Territories are facilitated by the collusion between the Palestinian Authority and Israeli political and security officials and businesspeople, so the continuation of the status quo constitutes a lucrative industry. The result of this reality is high levels of inequality in the West Bank, with on the one hand, very high levels of poverty and unemployment, while on the other

[3] Abraham Accords Declaration: Establishing diplomatic relations between Israel and neighboring countries in the region namely: UAE, Bahrain, Morocco and Sudan and encourage economic cooperation (US state department, 2022).

[4] Oslo Accords: Declaration of Principles on Interim Self-Government Arrangements (1993 Oslo Accords). It is a bilateral agreement signed in Washington following negotiations between Israel and the PLO (Asali Nuseibeh, 2015).

hand it is estimated that the total GDP of members of The Palestinian Development and Investment Company (PADICO)—founded by a small group of Palestinian businesspeople headed by the Al-Masri family—is roughly US$20 billion (Dana, 2020).

One of the most alarming results of this combination of colonization with neoliberal policy is the destruction of the Palestinian agricultural sector. Israel's settler colonialism and land appropriation (in the form of "settlements") in the West Bank almost doubled after the Oslo Accords, leaving less and less land for Palestinians. At the same time, Palestinian agricultural produce also shifted from key staples to luxury products to fill the gaps in Israeli exports and prevent competition with Israeli goods (Seidel, 2019). Added to that, Palestinian farmers are facing land confiscation threats from the Palestinian Authority, in order to build industrial zones, with international sponsorship and backing, these zones will eventually turn farmers into minimum wage labourers, and will eventually serve the interests of Israeli and political-elite Palestinian interests (Seidel, 2019).

Palestinians have been fragmented into small areas under the Israeli authority; leading to each one of those fragments—West Bank, Gaza, East Jerusalem and Palestinian community within Israel's 1948 borders, to have different economic and political realities. This means that each area has become a unique case and requires a specific analysis. In the case of East Jerusalem, Israel's policies (especially those that deny access to Jerusalem for the majority of Palestinians that do not have Jerusalem residence) have transformed this part of the city from a major Palestinian metropolitan center into an Israeli periphery that has high levels of poverty, limited economic opportunities and precariousness (see Shtern, 2018).[5]

Job Informality and Precariousness

> People are stumbling, between the need to work and get money, because the economic situation in Egypt is hard, and between trying to work for specific career goals, and also some don't even have any goals and don't know much about themselves.[6]

[5] Interviewee Faris (Pseudonym), 27, Muslim, Male, Shop Owner, Jerusalem.

[6] Interviewee Salim (Pseudonym), 19, Muslim, Male, Political Science Student, Cairo.

In 2011, Guy Standing published his much-debated book about labour precarity, which he defined as the combination of temporary insecure unstable jobs with inadequate income, insufficient work-related protections and benefits—such as pensions, paid sick leave, paid holidays, maternity leave and medical coverage. The "precariat" has no secure "occupational identity" (Standing, 2014: 969). They have to do a lot of work applying for jobs, constantly networking, filling forms, so they are constantly exploited outside and inside the workplace (Standing, 2014). They don't use their full educational qualifications and are mostly working in jobs that are not building their skills and enabling them to build a career and advance their opportunities (ibid.).

> My sister worked in so many fields unrelated to what she studied. She worked at Vodaphone, she worked as an English teacher. Now she's a journalist and she also works part time in an electricity company.[7]

> I see graduates spend a year or two being lost, not finding jobs, not knowing what to do... my brother spent two years not working, and finding random jobs unrelated to his degree, because there aren't many job opportunities and there are so many accountants in Jordan and most of them are unemployed.[8]

> If you see an honest person reaching places in life and he is not a thief or a crook, know that this person has been slapped around, humiliated and dragged in mud to reach where he has reached... I worked as a street vendor for three months to collect money for my wedding... after graduation from university I had a very low salary, you can't get married on that.[9]

> I was told that whatever university degree I graduate with I will not be able to work with it. Unfortunately, this is the system in Egypt. So many of my relatives including my own mother, they have a degree in one thing and then they are forced to work in another.[10]

[7] Interviewee Narmeen (Pseudonym), 18, Muslim, Female, Political Science Student, Cairo.

[8] Interviewee Aziz (Pseudonym), 24, Muslim, Male, Visual Communications Student, Amman.

[9] Interviewee Hakeem (Pseudonym), 34, Muslim, Male, Head of NGO, Jerusalem.

[10] Interviewee Ahlam (Pseudonym), 18, Muslim, Female, Computer Science Student, Cairo.

Standing (2014) also explains that this is the first generation in history that is expected to have a level of education that is greater than the labor they are expected to perform. Although Standing is not a MENA region specialist, and his work has been criticized by Munck (2013) as being Eurocentric, the description of the precariat and their relation to labour, as defined by Standing, can be witnessed among Arab youth, especially in the four contexts we are studying.

> I work eight hours a day and many many many times for eleven hours or more, for the minimum hourly wage… many times older people come and put their CVs, and there as so many needy people, they beg us for a job, but they take only younger people, maybe because they can enslave us, and make us work harder as young people. I really don't know.[11]

> There's no chance here. If you're an architect you're unemployed. If you're an architect you're an Uber driver.[12]

> I have a friend who graduate in accounting at Al Quds university, and then he went and did a Masters degree in Turkey. When he came back to the country he was applying everywhere, at banks and at universities and nobody accepted him. ….. he had big ambitions, because he had a Masters degree. The last time I saw him, he said to me, can you check if they have jobs at (a fast-food chain), will they take me there? In the end he was desperate, after looking and looking and looking for jobs.[13]

The story of precariousness can be seen in our four contexts. Job informality became the norm for the majority of educated youth, with unstable payments, limited access to social security, health insurance and union representation. The ILO (2021) estimates that 64% of total employment in the Arab region is informal, with unstable and poor working conditions. In Egypt, for example, prior to the uprising, the cohort of young people were the most educated in Egypt's history, with high job expectations, yet faced with a precarious employment situation, with the majority of them working in insecure circumstances with no formal contract (Silver, 2007). Bayat (2015: 34) has identified this social class group as

[11] Interviewee Rima (Pseudonym), 23, Muslim, Female, Employee at McDonalds, Jerusalem.

[12] Interviewee Ayman (Pseudonym), 21, Muslim, Male, Architecture Student, Amman.

[13] Interviewee Ahmad (Pseudonym), 23, Muslim, Male, Accounting Student, Jerusalem.

the "middle-class poor." This group, Bayat explains, holds educational capital and has high expectations, with a propensity for political engagement; it longs for middle-class life standards. However, economic and political policies have pushed these groups to work in precarious insecure jobs and live-in informal slums, so they become unable to lead the kind of life that they desire, leading them to exhibit outrage.

> The financial situation is a big problem, nobody is chasing their dreams, we need money to survive, only to survive. To buy a high-quality book, you will have to pay 250 Egyptian pounds, why would you do that? you can use this money to eat for a couple of days.[14]

Tadamun (2018) has mapped out the neighborhoods within greater Cairo to show the levels of education, unemployment and poverty, and they have noted that there are great disparities even within neighborhoods. They have also noted that even university education does not provide equal dividends in the different neighborhoods in Cairo, and not everyone is equally able to translate their higher education into employment opportunities, so the spaces within the city offer differing opportunities for people residing in different areas of the city. The mapping that Tadamun (2018) does shows that the relationship between education, employment and poverty is not straightforward. Gaining higher education does not automatically lead to employment.

> I'm from a working-class family and I'm now getting top education, so that puts me in the mid-level. Any person with a top education if they are from a higher-class family, that's it, they are set, and they don't need to worry about their career, they will eventually find something anywhere, but for me it's different.[15]

There are 26 neighborhoods in Cairo that they call "doubly advantaged" in terms of having high levels, in both higher education and employment levels; these neighborhoods are all located in either central Cairo, or in the new and expensive neighborhoods to the east and west (Tadamun, 2018). In these neighborhoods, employment is formal and stable, and the income of the residents is more than double that of neighborhoods with

[14] Interviewee Salim (Pseudonym), 19, Muslim, Male, Political Science Student, Cairo.

[15] Interviewee Salim (Pseudonym), 19, Muslim, Male, Political Science Student, Cairo.

lower levels of higher educational attainment (Tadamun, 2018). On the other hand, there are 66 neighborhoods that have high levels of higher education, but with high unemployment rates. Tadamun thinks that these are the middle-class neighborhoods that have youth with higher job expectations, but lower opportunities, so young people opt to wait until "suitable" formal and more stable opportunities are available. They also note that high unemployment rates are scattered across neighborhoods in Greater Cairo, but there are more urban neighborhoods (131) with high unemployment levels than rural neighborhoods (86). The neighborhoods with the "double disadvantage", which is low levels of higher education and low employment are scattered away from the center of Cairo to the south and north.

Poverty salaries were also a common theme among youth in the four cities. The minimum wage in Jerusalem is higher, but the costs of living are higher as well.

> The phrase "save money" it's not applied in Tunisia, because the salaries are so low….eighty per cent of this country is surviving, it's not living.[16]

> I get paid less than the older more established actors, and my effort is worth more than what I get paid, but to be able to pay for my expenses and transportation and various things, one needs to be enslaved a little. One needs to sacrifice, in order to reach what you desire in the end. This is the daily struggle we go through.[17]

> My brother is not happy with his job. He works very very hard and his salary is so low, but because there aren't many other options, he's stuck at this job that he hates.[18]

Poverty salaries make the transition to adulthood very difficult for youth, affecting their self-confidence, mental health and ability to establish independence. In the "Power 2 Youth" survey, between 66.8% of young people in Palestine, and 85.2% in Tunisia still lived in their family home (Calder et al., 2017). The precarious employment situation for youth in Egypt also means delayed transition from the family house and delayed marriage and independence. Assad and Barsoum (2009) reported that

[16] Interviewee Khalid (Pseudonym), 23, Muslim, Male, Unemployed, Tunis.

[17] Interviewee Khalil (Pseudonym), 25, Muslim, Male, Dancer/Actor, Tunis.

[18] Interviewee Salma (Pseudonym), 21, Muslim, Female, Marketing Student, Jerusalem.

in 2006 in Egypt, 57% of men in urban areas were not married by the age of twenty-nine, and 22% were unmarried by thirty-four, due to lack of income and housing, which is a serious cause of frustration.

> When I tell people that half my salary is going to rent and then the rest goes between the car and the groceries, they say to me immediately why don't you move back with your parents, and keep those 300JD, but I say no I'd rather not, I'd rather be independent.[19]

> My parents are funding me and It's so frustrating, but there's nothing I can do about it. You have two choices you either end up broke and homeless or you just, you know, you accept the help.[20]

> My fiancé and I have been engaged for over a year, but we are unable to save money for a small party... everything is expensive now in Tunis.[21]

It's not only in the public sphere that youth are excluded from, in the private sphere also youth expressed a deep sense of exclusion. Survey data reveals that large numbers of Arab youth feel that they are not taken seriously by their families, forcing them to abide by restrictive family rules that clash with their new global identities that they are developing over the social media (Calder et al., 2017).[22]

> Even people who work are still living with their parents. The salaries here are so low you cannot afford to live alone and pay for rent. I don't like being dependent on my family, that's why I want to look for decent work abroad.[23]

> Parents control of their children's lives is a big problem. They decide for us what to study and what to do. My sister had a scholarship to study abroad and my father refused to send her and now he regrets denying her this opportunity.... The parents they take these decisions without consulting

[19] Interviewee Mahmoud (Pseudonym), 25, Muslim, Male, Marketing Employee, Amman.

[20] Interviewee Aziz (Pseudonym), 24, Muslim, Male, Visual Communications Student, Amman.

[21] Interviewee Raneen (Pseudonym), 25, Muslim, Female, Pharmacist Assistant, Tunis.

[22] Interviewee Nasim (Pseudonym), 25, Muslim, Male, MBA Student, Jerusalem.

[23] Interviewee Omar (Pseudonym), 19, Muslim, Male, Visual Communications Student, Amman.

> their children. They just say no means no, and that's a big problem that many youth face.[24]

Palestinian Arab youth living in East Jerusalem face an added layer of discrimination. Since 1967, Israel has systematically closed Palestinian enterprises and businesses in East Jerusalem, under the claim that they are affiliated with the Palestinian Authority or due to "security claims", which has resulted in the weakening of the East Jerusalem economy (Asali Nuseibeh, 2019; UNCTAD, 2013). Other measures that have suffocated the Palestinian economy in Jerusalem include separating it from the rest of the Palestinian territories, preventing the rest of the Palestinian community from commuting to the Old City of Jerusalem without permits—which are increasingly difficult to obtain via the rigid bureaucratic system. Palestinian Jerusalemites looking for work opportunities can either cross the Israeli checkpoints daily to pursue work in the West bank, navigate the weak labour market in East Jerusalem, or explore options in Israeli cities or in the Israeli Western side of the city (Asali Nuseibeh, 2019). Palestinian Jerusalemites' opportunities are limited in both the Israeli and the Palestinian labour markets (ibid.). Pursuing employment in other Palestinian cities is difficult as the opportunities are limited, the salaries are low relative to the expenses people have in Jerusalem and people need to cross checkpoints on a daily basis, which subjects them to long travel hours, transportation expenses, and the constant risk of humiliation and intimidation by Israeli soldiers (Asali Nuseibeh, 2015, 2019).

> I have some friends who worked in Ramallah and it was so hard for them. They have to take several buses, and they get paid so little and in the end most their salary is spent on transportation, so they didn't feel it's worth it, so they left and started looking for work in Jerusalem.[25]

> For us Jerusalemites when you graduate from a Palestinian university you only find a job as a waitress or a cashier in an Israeli shop. You end up not working in the field you studied, and only for the minimum wage in Israel, or you work in Ramallah in your field of study, but also for the minimum wage of Ramallah, which is even less, and you spend a lot of time on

[24] Interviewee Narmeen (Pseudonym), 18, Muslim, Female, Political Science Student, Cairo.

[25] Interviewee Salma (Pseudonym), 21, Muslim, Female, Marketing Student, Jerusalem.

checkpoints and waste all your money on transportation, so people don't work in their field of study, you find them in restaurants and hotels.[26]

When Palestinians pursue work in the Israeli labour market, they face several barriers; the language and cultural barrier, lack of social networks, along with sometimes prejudice and racism from employers, colleagues and customers (Weissberg, 2014). A major problem they face is that some degrees from Palestinian universities are not recognized, so a postgraduate student from Al Quds University can end up as a daily labourer with precarious hazardous working conditions in the Israeli labour market, or in fields that do not require their level of education.

I was hoping that I would find a company in Jerusalem to work in, or a foreign NGO. I couldn't find a company on the Arab side, All the companies are on the Israeli side, and they all requires that you have strong Hebrew. As for NGO's, they all ask for experience, some ask for voluntary work..... but even when you volunteer it does not always translate to an actual job after.[27]

I would have never imagined myself to keep working at (a fast-food chain) to this day. I had huge ambition, and I wanted to be a famous journalist, but now I'm so disappointed. I thought that I would be something great and I worked hard for four years at the university for nothing. This is really upsetting, but this is what happened.[28]

Racism and abuse at work can also be an issue for Palestinians, resulting in them quitting their jobs or getting laid off, and in some extreme cases being taken to prison, as a result of an altercation with an Israeli customer. It can also be unsafe for them during times when there is escalation of violence to enter Israeli neighborhoods.

Working for Israelis you face a lot of problems and a lot of racism from customers saying "I don't want an Arab serving me", especially in Israeli

[26] Interviewee Salma (Pseudonym), 21, Muslim, Female, Marketing Student, Jerusalem.

[27] Interviewee Salma (Pseudonym), 21, Muslim, Female, Marketing Student, Jerusalem.

[28] Interviewee Rima (Pseudonym), 23, Muslim, Female, Employee at McDonalds, Jerusalem.

> settlements, but you get paid better there, in comparison to Arab areas, and so you swallow a lot of things, in order to get a better pay.[29]

Precariousness and discrimination in the labour market are not only defined by ethnicity and social class, gender also is a major issue affecting people's opportunities in the labour market.

Gender Discrimination and Harassment

The participation of Arab women in the labour force is among the lowest in the world (Global Gender Gap Report, 2021). Although fertility rates have declined and women's educational attainment in the region has increased, and even in several Arab countries has surpassed that of men, their participation in the labour market is still lagging behind. Assaad et al. (2020) argue that when we look at the gender gap in the labour market, we should shift our gaze from the supply-side factors to the demand-side issues of the labour market. The slowdown in public sector hiring, which used to be a main employer in Arab countries, especially for educated women, has left women with limited opportunities. They explain how this has not been compensated for with formal private sector jobs, leaving unemployment as the remaining "option" for women.

> My sister she would tell me how hard it was working in a private company. She would work from 8 in the morning to 8 in the evening, and would get so much money deducted if she missed a few hours. In addition to the psychological stress at work. That also, in addition to people who had connections in her company and received privileges that she didn't. She was so happy to shift to a governmental Job where the hours are reasonable and the workload and stress are reasonable.[30]

Although the overall numbers reflect women's exclusion in the labour market, each country and even each city and neighborhood has its own unique story, and people experience disadvantage along intersectional ethnic, gender and social class lines. As we have discussed previously, people experience the city differently and the residential area or

[29] Interviewee Adam (Pseudonym), 21, Muslim, Male, Employee at a Shop, Jerusalem.

[30] Interviewee Narmeen (Pseudonym), 18, Muslim, Female, Political Science Student, Cairo.

neighborhood can determine a person's life opportunities. Looking at Greater Cairo Region, for example, one can see vast differences between neighborhoods in terms of education, employment and poverty levels. Women's opportunities also vary vastly between neighborhoods in Cairo; for example, women's labour force participation in Cairo, can range from less than 2% in some neighborhoods to over 70% in others (Singerman & Higgins, 2018), which means that theories that blame the lack of job opportunities for Arab women on Arab and Islamic culture are inadequate. It is true that gender norms do affect women's opportunities and "choices" in the labour market, but there are also several economic and political factors that contribute to their weak participation. Labour force participation also only tells half the story, the type of jobs available is also important, whether they are formal stable jobs or temporary precarious ones. Singerman and Higgins (2018) tell us that in seventy neighborhoods in Egypt three-quarters of women's jobs are precarious and temporary. This intersects with lower levels of education and higher levels of poverty, meaning that uneducated women living in poverty are more likely to take exploitative precarious jobs out of desperation.

Palestinian Jerusalemite women face multiple layers of structural discrimination in finding employment, as a result of the combined effect of gender discrimination, neoliberalism and Occupation (Asali Nuseibeh, 2019). The participation of Palestinian women at peak working ages (25–64) in Jerusalem stands at 23%, in comparison with 81% of Israeli women of the same age cohort (Korach & Choshen, 2021). These numbers are also decepetive, as Palestinian women working in the informal sector, whether as domestic workers in the Israeli neighborhoods, Palestinian neighborhoods in Jerusalem or as employees in the West Bank are not counted in these figures.

> My sister she studied pharmacy at Al Quds university and of course it's not recognized (by Israel), so she's working as a teacher now. She used to work at a pharmaceutical company in Ramallah, but it was too much with the checkpoints. She would finish at four then spend three hours at the checkpoint and all for 500$ a month. It wasn't worth it.[31]

[31] Interviewee Mazen (Pseudonym), 25, Muslim, Male, Accounting Student, Jerusalem.

The combination of poor employment prospects, limited demand in the labour market, gender norms, societal pressures, along with limited child-care and inadequate labour laws protecting women's rights at work affect women's chances in the labour market (Asali Nuseibeh, 2019). Although the case of Palestinian women in Jerusalem is more severe due to the additional layer of discrimination, Arab women in Tunisia, Jordan and Egypt also face discrimination in the labour market along intersectional lines of gender and social class.

> In the lab where I work the majority are men and they are all older and get paid more, but even recently a new guy came after me and he's younger than me and his salary is higher. I don't know what the reason is. He managed to negotiate a better deal...... My salary is enough for me to eat and pay for transportation and a few expenses. It doesn't enable me to live alone.[32]

Another issue limiting the opportunities of women in the labour market is that they are clustered in "female-dominated" professions and so they compete with each other in a smaller pool of jobs. Women are pushed, directly and indirectly since primary and secondary school towards specific professions that are more "acceptable" for women, which is why they are also found clustered in the fields of education and social sciences at institutions of higher education and later in the labour market.

> I wanted to study mechanical engineering and I was told by my mother and other people also that I cannot join this field, because I am a girl and they would call me a mechanic... a career advisor also told me that this is not a profession for a woman, and they would choose a man for it, but when I kept saying to him I can do what a man can do, he said to me, also what would your husband say about you if this was your profession.[33]

Young women also face sexual harassment on the street and in the work environment. This is a major issue affecting and essentially limiting the life prospects of women in the city. It was mentioned by some of the interviewees as the most challenging aspect of living in their cities. Public sexual

[32] Interviewee Sina (Pseudonym), 23, Muslim, Female, Dental Lab Technician, Tunis.

[33] Interviewee Ahlam (Pseudonym), 18, Muslim, Female, Computer Science Student, Cairo.

harassment seems to be more of an urban phenomenon (Arab Barometer, 2019). In Egypt, public sexual harassment is more prevalent than in Palestine, Jordan and Tunisia. Sixty-two percent of the Egyptian women sampled in the Arab Barometer (2019) survey have experienced verbal public sexual harassment.

> I am a person who likes to walk and one of the things that makes it not desirable is that there's a lot of harassment on the street. While I walk, someone says something, or someone does something, so it's not the most pleasant thing.[34]
>
> I don't feel safe especially at night. Many times, even when I'm close to my house a car comes and parks and they stare at me until I leave the car, and even once a car kept coming close towards me to stop me from getting into my building, and once a guy followed me into my building, so many things like this happen to me.[35]
>
> I dress in a way that makes you feel that the person infront of you is not a woman, I give the impression that I can attack you if I wanted to. Of course, I don't like to dress this way, but I feel that the environment forces me to try and show that I am stronger than I actually am.[36]
>
> Whenever I see a group of men gathered together, I become very alert, I don't feel safe.[37]

Gender-based discrimination and sexual harassment in the workplace combined with lack of clear anti-harassment policies and an environment that shames the victim affects Arab women's recruitment, retention and promotion in the labour market (CIBIL, 2020). For example, in Egypt, sexual harassment in public is criminalized in its Penal Code, but the labor law lacks alignment and it is not criminalized it in the workplace (ibid.). Also, it is not enough to only have a legal framework against sexual harassment in the workplace; it needs to be implemented, and for this there needs to be educational and awareness programs that work on creating a safer workplace culture (ibid.). As long as these issues are not

[34] Interviewee Haya (Pseudonym), 19, Christian, Female, Economics Student, Cairo.

[35] Interviewee Samia (Pseudonym), 20, Muslim, Female, Art Student Amman.

[36] Interviewee Ahlam (Pseudonym), 18, Muslim, Female, Computer Science Student, Cairo.

[37] Interviewee Narmeen (Pseudonym), 18, Muslim, Female, Political Science Student, Cairo.

addressed properly on all levels, governmental, private sector employers, civil society and media, women will remain disadvantaged in the labour market.

> Everything a woman does gets sexualized, meaning if she smiles it means something. If she jokes too much, it means she's not a good woman.... These things don't happen to men.[38]

> Sometimes, as a girl, you sometimes worry when you want to go to an office, you think, what sort of people work at this office? will it be comfortable? will it be not comfortable. I don't think guys worry about these issues.[39]

> There was a guy with me who treated me in a disgusting way that was not comfortable at all.....He tried to touch me, so I hit him with the chair and broke the desk and went up to the management office and reported him and then I quit....he's still working there.[40]

Corruption and Nepotism

Lack of opportunities among Arab youth from weak economies are made worse by corruption, which permeates both political and economic institutions in the Arab world. Corruption and *Wasta*[41] are major issues corroding Arab institutions and affecting the opportunities of Arab youth. According to the Arab Opinion Index (2018), there's a clear consensus among the Arab public that corruption is widespread, with 76% of their sample agreeing that there is widespread corruption. The Arab Barometer Report across fifteen Arab countries shows an even higher perception regarding levels of corruption in state institutions at 84% (Al-Shami, 2019).

The abolition of Ben Ali's regime did not end the deep economic and social malaise of Tunisia. The International Crisis Group (ICG) (2017)

[38] Interviewee Narmeen (Pseudonym), 18, Muslim, Female, Political Science Student, Cairo.

[39] Interviewee Salma (Pseudonym), 21, Muslim, Female, Marketing Student, Jerusalem.

[40] Interviewee Farah (Pseudonym), 24, Muslim, Female, Shop Employee, Jerusalem.

[41] Wasta comes from the Arabic root Wasat, which translates to middle in English and it refers to intermediaries and connections that help one secure a position at an institution, or get over a bureaucratic process. It's the Arabic word for nepotism.

published a report on the regional socio-political challenges and corruption in Tunisia, arguing that the unrest in Tunisia was caused by the confrontation between the economic elite of northern Tunisia and the emergent elites of the south. Gallien and Hammami (2017) however warn of this kind analysis of the unrest in Tunisia as too simplistic. They explain that the economic system of Tunisia is very complex, with a clear interdependence between the formal and informal sectors. They have also explained how the North–South divide is not as clear as the ICG make it sound. The social and economic networks in southern Tunisia are not limited to just one group of ambitious smugglers and paid rioters, and its northern elites cannot be reduced to a single economic mafia protecting its privileges. The south has an urban–rural divide and some large-scale smugglers with different connections to Tunis, Tripoli and Beijing, in addition to a large middle size group of smugglers who work independently.

Corruption and nepotism are also rife in Egypt, Jordan and Palestine. The results of the 2021 Corruption Perceptions Index show that countries that protect and respect civil and political liberties are more able to control corruption, because the fundamental rights of freedom of expression, and freedom of press are important in the fight against corruption (OCHA, 2022). One hundred and eighty countries get ranked using this index with a score from zero—meaning no corruption—to one hundred—meaning rife with corruption. Egypt has the highest level of corruption among our four countries; it was rated 33 in the 2021 index, putting it at 117 in the world order (Transparency International, 2022). Tunisia scored 44 in the index, putting it at 70 in the world order. Jordan's score was better at 48, putting it at 58 in the world order. Palestine was not rated and Israel scored 59 in the index, putting it at 36 in the world order (ibid.).

Arab youth are continuously frustrated by corruption in the economy. They believe that corruption and *Wasta* is a major obstacle affecting the region, with 77% of youth in the Arab Youth Survey (2021) expressing concern over corruption in their countries, as well as citing corruption and economic hardship as main reasons behind their desire to emigrate. In the Arab Barometer Report, 88% believed that one needs *Wasta* to obtain employment (Al-Shami, 2019).

> I applied for so many places, in companies, banks, NGO's. They don't even get back to you with a phone-call, nothing...... You need to have

> *Wasta* that's it, nothing else.. Also, the political parties, if you belong to one of them, if you are loyal to a certain person, they give a phone call and you get a job, so this is our biggest problem.[42]

> The thing I see most of here is *Wasta*. This is something not good in this country. You see someone smart and they finish university and get their degree and they don't find a job, because of some other guy who's the son of I don't know who.[43]

> I didn't want to study for several years at the university to then end up working as an employee at a bank, which is anyway impossible because of the *Wasta*.[44]

Employment policies that focus solely on the youth assuming that they are the problem are looking only at half the picture; for example focusing only on building human capital and preparing the youth to enter the labour market via increasing their capacities, so they are better equipped to enter the private sector is not enough. Policies need to be directed at fighting the current corrupt institutions within the labour market and creating more demand and more opportunities for youth. The Employment policy in the Arab world has shifted from a focus on supply of labour opportunities to ensuring there is an availability of adequate demand for labour. There was a shift from providing governmental employment opportunities, towards developing human capital and increasing employees' capacities to work in the private sector (Calder et al., 2017). However, youth are disillusioned with this policy, as they can see that personal capacities and education are less important when it comes to finding stable employment than a person's connection and parents' wealth.

Calder et al. (2017) argue that, it is true that there is a mismatch between the youth education and skills and what is needed in the labour market, but also one cannot ignore the fact that the labour market does not supply enough stable jobs for youth. An even bigger challenge than finding employment for them to open their own private businesses or start-ups. It is hard for Arab youth to start their own businesses in an environment where there is corruption, nepotism, over-regulation, high running costs, lack of financial support and inability to compete fairly with

[42] Interviewee Nasim (Pseudonym), 25, Muslim, Male, MBA Student, Jerusalem.

[43] Interviewee Omar (Pseudonym), 19, Muslim, Male, Visual Communications Student, Amman.

[44] Interviewee Khalil (Pseudonym), 25, Muslim, Male, Dancer/Actor, Tunis.

the economic elite who have privileges and connections. As well as facing the constant risk that their ideas will be copied, especially since there are no copyright protections.

> I'm scared of opening my own business, because first the rent is so high and also, I look at people around me who opened their businesses in Jerusalem and they either fail and close down, or they are still open, but are struggling and all the effort and money that gets poured into the business is way more than the benefit.[45]

> There is no support for youth who want to start business projects at all. But there is a lot of money reaching Jerusalem for projects like this to institutions, and everything gets stolen in the end .... Rent in Jerusalem is so so so expensive. It can reach up to 4000$ a month for a shop in Jerusalem. In addition to the cost of labor, the running costs, the taxes.[46]

> When I studied management, I used to think life is good and you can graduate and open your own business...I have a business plan, but I don't have any money and I'm scared of sharing my ideas with people. I have no protection. I know someone she had a great plan and she presented it to a group of people and her idea was stolen. We don't have any protections or copyrights.[47]

> I stopped trusting people. If someone comes and says to me, we are an institution that supports projects in Jerusalem; I will say to them you are liars. You are either doing this for the media and the buzz in the news, or you are here to steal my ideas or you are commercial and trying to make money.[48]

> They came to our university, and they were like give us business ideas. They give you maximum 500$ if your idea is good. They steal ideas they don't invest in young people.[49]

Calder et al. (2017) explain that having a policy that only focuses on developing youth's entrepreneurial skills and building their capital, without adjusting the structural realities of their political economies

[45] Interviewee Salma (Pseudonym), 21, Muslim, Female, Marketing Student, Jerusalem.

[46] Interviewee Faris (Pseudonym), 27, Muslim, Male, Shop Owner, Jerusalem.

[47] Interviewee Nasim (Pseudonym), 25, Muslim, Male, MBA Student, Jerusalem.

[48] Interviewee Nasim (Pseudonym), 25, Muslim, Male, MBA Student, Jerusalem.

[49] Interviewee Ahmad (Pseudonym), 23, Muslim, Male, Accounting Student, Jerusalem.

can be detrimental to youth. Encouraging youth to develop their entrepreneurial skills and borrow capital and start their own small businesses in an environment where these businesses cannot succeed, can cause the youth to be further in debt and harm their career progression further (Calder et al., 2017). Debt is a major issue affecting youth in Arab countries and its worsening rapidly over time. The Arab Youth Survey (2021) reveals that 35% of its respondents were in debt in 2020, in comparison with 15% in 2015. The problem is worse in Jordan and Palestine, with 70% of young respondents in Jordan, 65% in Palestine, 28% in Tunisia and 20% in Egypt reporting to be in debt (ibid.). Herrera (2017) argues along the same lines in her discussion on Arab youth employment prospects. She says that entrepreneurship skills, drive and ambition are positive qualities, but it is unfair to propagate the myth that anyone with an idea, grit and determination can be a successful entrepreneur in the Arab world.

> I had a start-up project, but I faced several problems......The economic situation in Egypt is not that great and I needed a budget for my business to be sustainable. I also lacked the experience, although I'm taking business entrepreneurship now, but I still lack the skills.[50]

Herrera (2017) explains that big companies that portray themselves as the heart of US entrepreneurship, have benefited from the US government-funded investments. Arab governments and businesses should prioritize investing in youth businesses and entrepreneurial ideas, as well as creating employment opportunities, rather than leading the youth down a road of borrowing and debilitating debt. Paciello and Pioppi (2017) also think that making youth employment problems purely a matter of individual characteristics, or deficiencies of the education system, obscure the structural factors of the political economy, which are behind the spread of precariousness and insecurity among youth. Counter to realities in OECD countries, education in Egypt, Tunisia, Jordan and Palestine is not a guarantee against unemployment, on the contrary, unemployment tends to increase with education, especially in Egypt. Assaad and Krafft (2015) argue that, in the case of Egypt, high unemployment affects the highly educated as well, who happen to be coming from middle-class groups, as they have the financial ability to remain unemployed until a suitable

[50] Interviewee Salim (Pseudonym), 19, Muslim, Male, Political Science Student, Cairo.

job becomes available. This is supported by Barsoum (2014) who says that in her research sample, 30% of the unemployed university graduates felt that the job opportunities available did not match their qualifications, which means that there is a problem with the supply in the market for educated workers. The major obstacle to youth employment in those Arab countries is the structural flaws in those economies, and the lack of good quality job opportunities that provide living wages, benefits and security. Policies that focus solely on increasing the quality of individuals' skills and qualifications can only raise expectations and intensify competition for scarce employment opportunities and, consequently, lead to feelings of frustration and marginalization (Calder et al., 2017).

> There are these commercial groups that advertise and say they want to support you in your project idea, and then when you go to them you realize that they are just making you pay to take courses and improve your skills in business, but you already have those skills and the degrees. You just want the investment, not more training.[51]

Migration

Economic problems and socio-political instability continue to fuel youths' desire to emigrate. Around 42% of Arab youth consider emigration, with their most desired destinations being Canada, United States, Germany and UAE (Arab Youth Survey, 2021). The number-one reason they stated for considering emigration was their economic situation (ibid.).

> My older brother studied business in Irbid, and then when he came back he looked for jobs everywhere; he couldn't find anything. Then he found a job in Tel Aviv and he worked for a year, then the company closed down and today he's a receptionist at a hotel... I don't think he will stay in the country. He's at a dead-end job, it won't get him anywhere.[52]

The ILO (2021) estimates that, prior to the COVID-19 pandemic, the Arab region needed to create an estimated 33.3 million new jobs between 2020 and 2030 to reduce the unemployment rate to 5%.

[51] Interviewee Nasim (Pseudonym), 25, Muslim, Male, MBA Student, Jerusalem.

[52] Interviewee Rima (Pseudonym), 23, Muslim, Female, Employee at McDonalds, Jerusalem.

> Me for example, when I graduated, I spent a long time without work, maybe five, six years, meaning it's too much… I was looking for jobs and just trying to fill my free time, so that I don't just sit doing nothing.[53]

This interviewee is one voice of millions in Tunisia. It takes an average of six years for university graduates in Tunisia to find a stable job, and by the age of 35, half of all university graduates in the country remain unemployed (Angel-Urdinola et al., 2015). It is true that the state of the Tunisian economy is the product of years of the corrupt Ben Ali regime, but the post-revolution governments were also inefficient in tackling the economic problems. They focused more on political reform and internal dispute deferring the much-needed economic reform, including fighting corruption, reforming customs, streamlining the bureaucracy and tax reform (Yerkes & Ben Yahmed, 2018). In Jordan, youth's refusal to accept low-wage employment highlights the mismatch between their levels of education/training, and the available low wages on offer in the economy, which has led to high levels of 'voluntary unemployment' (Taghdisi-Rad, 2012). This situation has also pushed Jordanians to seek jobs abroad. The Jordanian economy relies on remittances from Jordanians working in other countries, 10% of the Jordanian population lives and works in the Gulf States (Brown et. al. 2014).

> I know for a fact that when I graduate from university, I'm not going to find a job, it's a fact. If I want to be anything in this country, it will have to be after I establish myself abroad, but to start here it's impossible.[54]
> Our lecturers at the university they make fun of us; they say even the best of you won't find a job imagine the bad ones… basically most people even my family they say to me you won't find a job, so it's a big concern for me. They say every year so many architects are graduating, and there are already so many of us unemployed.[55]

Talented and qualified Arabs tend to migrate outside their home country and the majority of those studying abroad tend to never go back, mainly because of better opportunities, resources, freedoms and political

[53] Interviewee Yumna (Pseudonym), 32, Muslim, Female, NGO Employee, Tunis.

[54] Interviewee Ayman (Pseudonym), 21, Muslim, Male, Architecture Student, Amman.

[55] Interviewee Samah (Pseudonym), 22, Muslim, Female, Architecture Student, Amman.

stability abroad. Palestine, Tunisia, Egypt, Jordan, Syria, Morocco and Lebanon suffer the most from the emigration of skilled talent, which is called the "brain drain". Every year over 3,000 IT engineers leave Tunisia for Europe, where wages can be up to 2.5 times higher (Field, 2020). Also, 75% of Egyptian emigrants have a university degree (Al Ghwell, 2018). This "brain drain" costs the Arab world an annual $2Billion (ibid.).

> We have a lot of brilliant students, who have good education and have a lot of potential. Engineers, doctors, PhD's.... the major problem is that when they are good, they want to get out of the country, because here they don't get money for it. They don't have the chance to have a career, so they go out. We are exporting so many good brains from Tunisia to other countries and they are taking them.[56]

Education and skill are not the only criteria leading youth to leave Arab countries. At the other end of the scale, lack of opportunities can lead to radicalization and emigration for political reasons. For instance deprivation and absence of hope for a better future has pushed many of Tunis's marginalized youths to join Jihadi groups in neighboring conflict zones. The Washington Institute for Near East studies released a report in 2018 on Tunisians participating in Jihadi activities in Iraq and Syria, which shows that a total of some 30,000 Tunisians either participated in these, or were mobilized, but were stopped and never made it to the conflict zones (Zelin & Walls, 2018). The report shows that among the motivations for Tunisians traveling to Iraq and Syria were disillusionment with post-revolution politics, pursuit of economic opportunity, and radicalization in prison. The report shows that Tunisians were the seventh highest foreign fighter group to have joined Jihadi groups in Syria and Iraq and in per capita terms, they were the third most represented group, behind Libyans and Saudis. Tunis the capital accounts for 36% of those fighters joining from Tunisia (ibid.).

Migration contributes to the slow pace of political change in the home countries (De Bel-Air, 2018). Highly skilled and educated Arab expatriates contribute to the reproduction of the socio-economic conditions and corrupt systems in their countries by choosing to "exit" rather than "voice" (Hirschman, 1970 cited in De Bel-Air, 2018: 69). Instead

[56] Interviewee Khalid (Pseudonym), 23, Muslim, Male, Unemployed, Tunis.

of staying in their countries and fighting the unjust systems of access to opportunities, resources and political power, they choose to exit their countries and look for opportunities elsewhere. Fargues (2017) has argued that it doesn't have to be either "exit" or "voice"; expatriation and political protest can work in synergy, they don't necessarily clash, as expatriation can be the result of failed political protest and mobilization processes, and on the other hand, protest can be fueled by expatriates who find inspiration in ideas and models gathered abroad. There is also the economic added value of remittances to their home countries, that can dampen the effect of the financial loss of the "brain drain". Tsourapas (2020) has noted how authoritarian states losing large numbers of emigrants face a dilemma; on the one hand, there are the economic benefits of remittances and the reducion of their unemployment rates, but on the other hand, they cannot control political dissenters and their activism abroad. These governments extraterritorial authoritarian practices can range from surveillance tactics, in order to monitor diaspora behavior, mobilizing pro-regime activists, and denying renewal of citizenship to, in more severe cases assassination, such as the case of Khashoggi (ibid.).

The major destinations of most Arab migrants, such as Europe and the Gulf States, now place increasingly tight restrictions on migration from Arab countries. Tragically this does not stop the migration of people fleeing war or political persecution, it only leads to irregular unsafe migration and the thriving of the people-smuggling industry. Since the Arab uprisings took place in 2011, the migration control deals that Europe had signed with previous Arab dictators to control movement, were no longer stopping migrants. The number of migrants and the number of deaths as a result of exploitative smuggling campaigns were actually increasing.

After the uprisings in Arab countries, Gulf States have also adopted restrictive policies against foreign residents, in fear of the risk of spillovers of regional conflicts, with deportations affecting nationals from countries that had witnessed regime change. For instance, in 2013, Saudi Arabi deported 300,000 Egyptians (De Bel-Air, 2018). Gulf governments started adopting policies that prioritized local employees, while limiting the access to immigrants from other Arab countries. Palestinian, Tunisian, Egyptian and Jordanian youth are therefore facing double push factors; the economic and political situation in their countries is pushing them to want to emigrate and pursue opportunities abroad; meanwhile, the traditional destinations for this emigration (European countries and Gulf

States) are restricting their access and pushing them out, resulting in what De Bel Air (2018: 52) defines as the "Blocked" youth.

Conclusion

> So you don't really know if the problem is from the young man himself, or the economic situation. It's a wicked problem as they call it, and it doesn't have one solution, so many solutions, and that requires a large number of people to gather together and solve it.[57]

As this young Egyptian man explains, the challenges in the labour market are entangled, and that requires a comprehensive approach to address the multi-dimensional problems plaguing Arab economies. This chapter has particularly focused on the political and economic factors that have shaped youth experiences in the labour market. The current labour markets conditions are unfavorable to Arab youth, who face the highest unemployment rates in the world (ILO, 2020). The gap in the Arab world between adult and youth unemployment is large, with new young entrants to the labour market being affected the most. In the 1950s and 1960s Arab youth in Jordan, Egypt and Tunisia benefited from the socialist government policies, which were manifested in free education followed by guaranteed employment in the government sector. However, these policies were unsustainable and created a "credentials fetish" society (Barsoum, 2017: 105), that primarily depended on government hiring, which was mainly available for politically significant groups and was mired with corruption. Economic privatization policies adopted from the 1970s onwards have created further challenges. They reduced government jobs, but did not manage to create formal stable employment in the private sector to compensate; instead, these policies have resulted in increasing job informality and precariousness. The new precariat as defined by Standing (2011) has no occupational identity. They keep jumping from one temporary job to another, they are forced to spend a lot of time applying for jobs and filling forms, to find themselves in jobs that do not meet their credentials, with low salaries and no benefit packages, keeping them in an impoverished state and dependent on their families.

[57] Interviewee Salim (Pseudonym),19, Muslim, Male, Political Science Student, Cairo.

This climate explains why there are high tendencies towards emigration, especially among youth from Egypt, Tunisia, Jordan and Palestine.

Going back to the days when the governments were major employers is not sustainable, but likewise continuing with privatization policies that are rife with corruption, and enjoyed mainly by political elite, with no serious competition, is also not sustainable either, and risks further eroding the middle class, and increasing inequality and unrest. Therefore, a new model should be identified that can combine both public and private economic cooperation. However, under authoritarian regimes, where corruption and monopolistic practices thrive, serious public–private synergies directed to serve society as whole are questioned. Due to the lack of democracy, stifled media, lack of freedom of speech, disempowered civil society, co-opted labour unions and absence of permission to demonstrate, governmental spending and policy remains unchallenged, even though they cater only to the needs of the few. An example of this is the massive budgets spent on SiSi's new city, rather than these funds being used to improve the state of the education and health sectors in Egypt. Private sector projects also remain in the hands of the few who are connected to the government, that is why the drive towards privatization did not really generate new job opportunities, as it only benefitted the few. Also, the focus on the real estate, tourism and services sectors, rather than export industries, have resulted in less job creation. However, whatever failed governmental policies are being adopted they go unchallenged in the current climate of media stifling.

In the Arab Youth Survey (2021), the second most stated reason for pushing them to seek emigration is educational opportunities. In the next chapter, we will explore the education sector in our four contexts, in terms of quality and accessibility from a "capabilities approach" perspective.

Bibliography

Abdelrahman, M. (2017). Policing neoliberalism in Egypt: The continuing rise of the 'securocratic' state. *Third World Quarterly, 38*(1), 185–202.

AlAzzawi, S., & Hlasny, V. (2020). *Vulnerable employment of Egyptian, Jordanian, and Tunisian youth: Trends and determinants*. United Nations University. https://www.wider.unu.edu/publication/vulnerable-employment-egyptian-jordanian-and-tunisian-youth. Accessed 16 May 2022.

Aleya-Sghaier, A. (2012). The tunisian revolution: The revolution of dignity. *Journal of the Middle East and Africa, 3*(1), 18–45.

Al Ghwell, H. (2018). Gulf states can teach the Arab world about brain gain. https://www.arabnews.com/node/1383526. Accessed 1 May 2022.

Al Shami, S. (2019). *Arab Barometer Report: Perceptions of corruption on the rise across MENA* https://blogs.worldbank.org/arabvoices/arab-barometer-report-perceptions-corruption-rise-across-mena. Accessed 22 May 2022.

Angel-Urdinola, D., Nucifora, A., & Robalino, D. (2015). Labor policy to promote good jobs in Tunisia revisiting labor regulation, social security, and active labor market programs. https://openknowledge.worldbank.org/bitstream/handle/10986/20604/928710PUB0Box3021027109781464802713.pdf?sequence=1&isAllowed=y. Accessed 22 May 2022.

Arab Barometer. (2019). Sexual harassment and domestic violence in the Middle East and North Africa. https://www.arabbarometer.org/wp-content/uploads/Sexual-Harassement-Domestic-Violence-Arab-Citizens-Public-Opinion-2019.pdf. Accessed 22 May 2022.

Arab Opinion Index. (2018). *2017–2018 Arab opinion index: Executive summary.* Arab Center. https://arabcenterdc.org/resource/2017-2018-arab-opinion-index-executive-summary/. Accessed 1 May 2022.

Arab Youth Survey. (2021). Hope for the future. https://arabyouthsurvey.com/wp-content/uploads/whitepaper/AYS-2021-Top-10-Findings-English.pdf. Accessed 1 May 2022.

Association for Civil Rights in Israel (ACRI). (2021). East Jerusalem facts and figures. https://www.english.acri.org.il/post/__283. Accessed 1 May 2022.

Assaad, R., & Barsoum, G. (2009). Rising expectations and diminishing opportunities for Egypt's young. In N. Dhillon & T. Yousef (Eds.), *Generation in waiting: The unfulfilled promise of young people in the Middle East* (pp. 67–94). Brookings Institution Press.

Assaad, R., Hendy, R., Lassassi, M., & Yassin, S. (2020). Explaining the MENA paradox: Rising educational attainment yet stagnant female labor force participation. *Demographic Research, 43*, 817–850.

Assaad, R. (2014). Making sense of Arab labor markets: The enduring legacy of dualism. *IZA Journal of Labor & Development, 3*(1), 1–25.

Assaad, R., & Krafft, C. (2015). Social background and attitudes of higher education Students and Graduates in Egypt. http://docplayer.net/153843587-Social-background-and-attitudes-of-higher-education-students-and-graduates-in-egypt-ragui-assaad-and-caroline-krafft-january-2015.html. Accessed 1 May 2022.

Asali Nuseibeh, R. (2019). Palestinian women teachers in East Jerusalem: Layers of discrimination in the labor market. *The Middle East Journal, 73*(2), 207–223.

Barsoum, G. (2015). 'Job opportunities for the youth': Competing and overlapping discourses on youth unemployment and work informality in Egypt. *Current Sociology, 64*(3), 430–446.

Barsoum, G. (2017). The allure of 'easy': Reflections on the learning experience in private higher education institutes in Egypt. *Compare: A Journal of Comparative and International Education, 47*(1), 105–117.

Barsoum, G., Ramadan, M., & Mostafa, M. (2014). *Labor market transitions of young women and men* (Work4Youth Publication Series. No. 16). http://www.ilo.org/wcmsp5/groups/public/@dgreports/@dcomm/documents/publication/wcms_247596.pdf. Accessed 22 May 2022.

Bayat, A. (2015). Plebeians of the Arab Spring. *Current Anthropology, 56*(11), 33–43.

Beinin, J. (1998). Palestine and Israel: Perils of a neoliberal, repressive "Pax Americana." *Social Justice, 25*(4 (74)), 20–39.

Ben Yahmed, Z., & Yerkes, S. (2018). *Tunisians' revolutionary goals remain unfulfilled*. https://carnegieendowment.org/2018/12/06/tunisians-revolutionary-goals-remain-unfulfilled-pub-77894. Accessed 15 March 2021.

Bogaert, K. (2013). Contextualizing the Arab Revolts: The politics behind three decades of neoliberalism in the Arab World. *Middle East Critique, 22*(3), 213–234.

Bonvin, J.-M., & Farvaque, N. (2007). A capa- bility approach to individualised and tailor- made activation. In R. van Berkel & B. Valkenburg (Eds.), *Making it personal: Individualising activation services in the EU* (pp. 45–66). The Policy Press.

Bonvin, J.-M., & Galster, D. (2010). Making them employable or capable? Social integration policy at the crossroads. In H.-U. Otto & H. Ziegler (Eds.), *Education, welfare and the capabilities approach* (pp. 71–84). Barbara Budrich Publishers.

Bournakis, I. (2021). *Income inequality convergence across Egyptian governorates—Economic Research Forum (ERF)*. https://theforum.erf.org.eg/2021/03/08/income-inequality-convergence-across-egyptian-governorates/. Accessed 22 May 2022.

Brown, R., Constant, L., Glick, P., & Grant, A. (2014). *Youth in Jordan transitions from education to employment*. https://www.rand.org/content/dam/rand/pubs/research_reports/RR500/RR556/RAND_RR556.pdf. Accessed 16 May 2022.

Bussi, M., & Dahmen, S. (2012). When ideas circulate. A walk across disciplines and different uses of the "capability approach". *Transfer: European Review of Labour and Research, 18*(1), 91–95.

Butler, D. (2020). *Egypt and the gulf*. https://www.chathamhouse.org/2020/04/egypt-and-gulf. Accessed 16 May 2022.

Calder, M., MacDonald, R., Mikhael, D., Murphy, E., & Phoenix, J. (2017). *Marginalization, young people in the South and East Mediterranean, and policy: An analysis of young people's experiences of marginalization across six*

SEM countries, and guidelines for policy-makers (Power2Youth Working Paper No.35).

CIBIL. (2020). *Combatting sexual harassment in the workplace: A daily struggle.* https://www.aub.edu.lb/cibl/news/Pages/Combatting_Sexual_Harassment_in_the_Workplace.aspx. Accessed 26 May 2022.

Clarno, A. (2017). *Neoliberal apartheid: Palestine/Israel and South Africa after 1993.* University of Chicago Press.

Dana, T. (2020). Crony capitalism in the Palestinian Authority: A deal among friends. *Third World Quarterly, 41*(2), 247–263.

De Bel-Air, F. (2018). 'Blocked' youth: The politics of migration from South and East Mediterranean countries before and after the Arab Uprisings. *The International Spectator, 53*(2), 52–73.

Dimova, R., & Stephan, K. (2020). Inequality of opportunity and (unequal) opportunities in the youth labour market: How is the Arab world different? *International Labour Review, 159*(2), 217–242.

Dridi, M. (2021). *Tunisia facing increasing poverty and regional inequalities.* https://carnegieendowment.org/sada/85654. Accessed 16 May 2022.

Elgendy, K., & Abaza, N. (2020). *Urbanization in the MENA region: A benefit or a curse?* https://mena.fes.de/press/e/urbanization-in-the-mena-region-a-benefit-or-a-curse. Accessed 16 May 2022.

El Gantri, S. (2019). *'The bird will not return to the cage:' An analysis of Tunisia's 2019 elections* | International Center for Transitional Justice. https://www.ictj.org/news/'-bird-will-not-return-cage'-analysis-tunisia's-2019-elections. Accessed 16 May 2022.

Fahmy, H. (2012). An initial perspective on "The winter of discontent": The root causes of the Egyptian revolution. *Social Research, 79*(2), 349–376.

Fargues, P. (2017). Mass migration and uprisings in Arab countries: An analytical framework. *Revue Internationale De Politique De Développement,* (7).

Gallien, M., & Hammami, M. D. (2017). *Corruption and reform in Tunisia: The dangers of an elitist analysis.* Jadaliyya. https://www.jadaliyya.com/Details/34323/Corruption-and-Reform-in-Tunisia-The-Dangers-of-an-Elitist-Analysis. Accessed 16 May 2022.

Global Gender Gap Report. (2021). Global gender gap report 2021 insight. *World Economic Forum.* https://www3.weforum.org/docs/WEF_GGGR_2021.pdf. Accessed 28 May 2022.

Gordon, N. (2008). *Israel's occupation.* University of California Press.

Hanieh, A. (2013). *Lineages of revolt.* Haymarket Books.

Hever, S. (2010). *The political economy of Israel's occupation.* Pluto Press.

Herrera, L. (2017). It's time to talk about youth in the Middle East as the precariat. *Middle East: Topics & Arguments, 9*, 35–44.

ILO (International Labour Organization). (2020). *'Global employment trends for youth 2020: Arab States'*. https://www.ilo.org/wcmsp5/groups/public/---dgreports/---dcomm/documents/briefingnote/wcms_737672.pdf. Accessed 16 May 2022.

ILO. (2021). *Towards a productive and inclusive path job creation in the Arab Region*. https://www.ilo.org/wcmsp5/groups/public/---arabstates/---ro-beirut/documents/publication/wcms_817042.pdf. Accessed 16 May 2022.

IMF. (2012). *Youth unemployment in the MENA region: Determinants and challenges*. https://www.imf.org/external/np/vc/2012/061312.htm?id=186569. Accessed 16 May 2022.

Independent Evaluation Group (IEG). (2009). *Egypt—Positive results from knowledge sharing and modest lending: An IEG country assistance evaluation 1999–2007*. World Bank. https://openknowledge.worldbank.org/handle/10986/13534. Accessed 16 May 2022.

International Crisis Group. (2017). Blocked transition: Corruption and regionalism in Tunisia. [online]. Available at: https://www.crisisgroup.org/middle-east-north-africa/north-africa/tunisia/177-blocked-transition-corruption-and-regionalism-tunisia. Accessed 16 May 2022.

International Trade Administration. QIZ Jordan. https://www.trade.gov/qiz-jordan. Accessed 16 May 2022.

Joya, A. (2017). Neoliberalism, the state and economic policy outcomes in the Post-Arab uprisings: The case of Egypt. *Mediterranean Politics, 22*(3), 339–361.

Kausch, K. (2009). Tunisia: The life of others. Project on freedom of Association in the Middle East and North Africa (FRIDE Working Paper, 85).

Khalidi, R., & Samour, S. (2011). Neoliberalism as liberation: The statehood program and the remaking of the Palestinian national movement. *Journal of Palestine Studies, 40*(2), 6–25.

Korach, M., Choshen, M. (2021). *Jerusalem facts and trends*. https://jerusaleminstitute.org.il/wp-content/uploads/2021/05/Pub_564_facts_and_trends_2021_eng.pdf. Accessed 16 May 2022.

Marshall, T. H. (1950). *Citizenship and social class* (T. Bottomore, Ed.). Pluto.

Malik, A., & Awadallah, B. (2013). The economics of the Arab spring. *World Development, 45*, 296–313.

Munck, R. (2013). The precariat: A view from the South. *Third World Quarterly, 34*(5), 747–762.

Murphy, E. (2017). A political economy of youth policy in Tunisia. *New Political Economy, 22*(6), 676–691.

Nguyen, C. (2022). A complete guide to public transportation in Benjing. Chibikin. https://www.chibikiu.com/blog/publictransport-in-beijing-guide#11_Beijing_public_transport_overview

OCHA. (2022). Corruption Perceptions Index 2021. https://reliefweb.int/report/world/corruption-perceptions-index-2021-enarru. Accessed 22 May 2022.

Paciello, M. C. (2011, May). Tunisia: Changes and challenges of political transition (Working Paper 2, MEDPRO Technical Report, No. 3) (pp. 1–26).

Paciello, M. C., & Pioppi. D. (2017, May). Youth in the South East Mediterranean Region and the need for a political economy approach (POWER2YOUTH Working Paper 37). IAI.

Pelham, N. (2011). Jordan's Balancing Act - MERIP. https://merip.org/2011/02/jordans-balancing-act/. Accessed 16 May 2022.

Royal Scientific Society. (2013). The future of Jordan's Qualified Industrial Zones (QIZs). https://library.fes.de/pdf-files/bueros/amman/10677.pdf. Accessed 16 May 2022.

Sayigh, Y. (2012). *Above the state: The officers' republic in Egypt* (The Carnegie Papers). (Carnegie Endowment for International Peace). http://carnegieendowment.org/files/officers_republic1.pdf. Accessed 16 May 2022.

Seidel, T. (2019). Neoliberal developments, national consciousness, and political economies of resistance in Palestine. *Interventions, 21*(5), 727–746.

Shtern, M. (2018). Towards 'ethno-national peripheralisation'? Economic dependency amidst political resistance in Palestinian East Jerusalem. *Urban Studies, 56*(6), 1129–1147.

Singerman, D., & Higgins, D. (2018). Gender, precarity, and inequality in Cairo's neighbourhoods—Urbanet. https://www.urbanet.info/gender-precarity-and-inequality-in-cairo/. Accessed 22 May 2022.

Sika, N. (2018). Neoliberalism, marginalization and the uncertainties of being young: The case of Egypt. *Mediterranean Politics, 24*(5), 545–567.

Sika, N. (2012, August). Youth political engagement in Egypt: From abstention to uprising. *British Journal of Middle Eastern Studies, 39*(2), 181–199.

Sika, N. (2016). Youth civic and political engagement in Egypt (Working paper no. 18). ISSN 2283-5792.

Silver, H. (2007). Social exclusion: Comparative analysis of Europe and Middle East Youth. *SSRN Electronic Journal.*

Sims, D. (2012). *Understanding Cairo. The logic of a city out of control* (2nd ed.). American University in Cairo Press.

Standing, G. (2011). *The precariat: The new dangerous class.* Bloomsbury Academic.

Standing, G. (2014). Understanding the precariat through labour and work. *Development and Change, 45*(5), 963–980.

Tadamun. (2018). Inequality of opportunity in Cairo: Space, higher education, and unemployment. http://www.tadamun.co/inequality-opportunity-cairo-space-higher-educationunemployment/?lang=en#.YciJgi2cY1I. Accessed 26 December 2021.

Taghdisi-Rad, S. (2012). Macroeconomic policies and employment in Jordan: Tackling the paradox ofjob-poor growth (Employment Working Paper No. 118). ILO.

Transparency International. (2022). 2021 Corruption Perceptions Index—Explore the results. https://www.transparency.org/en/cpi/2021. Accessed 16 May 2022.

Tsourapas, G. (2020). The long arm of the Arab state. *Ethnic and Racial Studies, 43*(2), 351–370.

United States Department of State. (2022). The Abraham Accords—United States Department of State. https://www.state.gov/the-abraham-accords/. Accessed 30 May 2022.

UNCTAD. (2013). *The Palestinian economy in East Jerusalem: Enduring annexation isolation and disintegration*. https://unctad.org/webflyer/palestinian-economy-east-jerusalem-enduring-annexation-isolation-and-disintegration. Accessed 16 May 2022.

Weissberg, H. (2014, March 31). Survey: 42% of employers prefer not to hire Arab men. *Haaretz*. www.haaretz.com/1.5341704. Accessed 16 May 2022.

Zelin, A., & Walls, J. (2018). *Tunisia's foreign fighters*. https://www.washingtoninstitute.org/policy-analysis/tunisias-foreign-fighters. Accessed 16 May 2022.

Zemni, S. (2017). The Tunisian revolution: Neoliberalism, urban contentious politics and the right to the city. *International Journal of Urban and Regional Research, 41*(1), 70–83.

Zureik, E. (2020). Settler colonialism, neoliberalism and cyber surveillance: The case of Israel. *Middle East Critique, 29*(2), 219–235.

CHAPTER 4

Education in the City

Introduction

The concepts of "The Learning City," "The Educating City," and "Child Friendly Cities" have called for the enhancement of the capabilities and the opportunities of the city residents (IAEC, 2004, 2020; UNESCO, 2020; UNICEF, 2022). These concepts call for building cities that are educated, green, healthy, smart, equitable, and child and youth friendly. They have all advocated for enhancing the educational rights of residents of the city, some have advocated for life-long learning, others have advocated for collaboration between cities to enhance the educational opportunities for residents and some have advocated for the participation of children and youth in the decision-making processes of educational provision.

These initiatives have focused mainly on what changes in the governance policies, structure, and resources are needed to improve educational opportunities and capabilities of the residents of the city. They have called for the mobilization of resources, political power and municipal commitment, to promote the enhancement of the quality of education from childhood through-out life (UNESCO, 2020).

Despite a number of projects implemented around the cities in the spirit of "The Learning City," "The Educating City" and "Child Friendly Cities", many children still dwell in polluted cities, poor living conditions

R. A. Nuseibeh, *Urban Youth Unemployment, Marginalization and Politics in MENA*, Middle East Today,
https://doi.org/10.1007/978-3-031-15301-3_4

and weak educational facilities that do not respect diversity. Arab cities, in particular, have yet to embrace the concept of the "Learning City." The high rate of urban expansion in Arab countries, due to rural–urban migration and high rates of natural population growth, has not been matched by a similar increase in health, education, housing and transportation services (Nour, 2013). There have been big achievements in the Arab world in terms of increasing literacy rates, increasing school enrollment and completion rates, closing gender gaps and in some countries even reversing gender gaps, but there are still issues with the quality and the relevance of the education offered.

> There should be things given at school that help us in general. ....General knowledge is very poor amongst the students, even at the university level the knowledge and education overall amongst the students is very poor... The students have nothing beyond what the book is offering.[1]

The focus in the Arab world is on the economic factors of education, and on reducing unemployment, while this is important, it neglects the enhancement of capabilities, democratic education, human rights, social engagement and active citizenship. The quality of the education received and equity of the provision of services remains a challenge in Arab cities to varying degrees. The drive towards neoliberal urbanism in Arab cities, which is manifested in the privatization of urban institutions and services, with extensive tax breaks and incentives for corporate development and investment, has seeped into educational institutions. Educational privatization drives cuts in government education budgets and pushes the integration of market logics into the fabric of urban public education (Means, 2014). Added to that is the continuous rhetoric that reduces the role of education to its market value. That is why several educational experts have pushed for the re-framing of educational objectives, to look at them through the lens of Amartya Sen's "capabilities approach" (Robeyns, 2006; Unterhalter, 2003; Walker, 2006). Most of the discussion on the education of Arab youth are framed through the lens of the human capital approach. The human capital approach is focused on the instrumental economic benefits of education, reducing unemployment, eradicating poverty, increasing the possibility of labour market participation, and the fact that modern labour markets depend on an educated

[1] Interviewee Ahmad (Pseudonym), 23, Muslim, Male, Accounting Student, Jerusalem.

workforce, etc. Robeyns (2006: 69) in her article on the three models of education: rights, capabilities and human capital looks at the three normative accounts that underlie educational policies. She explains that the human capital approach, although useful, is limited, as it is blind to cultural dimensions of societies, because people do not solely base their decisions on economic gain, and there are other religious, cultural, emotional and non-monetary motivations behind people's actions in societies (cited in Asali Nuseibeh, Forthcoming). Second, she explains that this approach can indirectly contribute to inequality, as it sees that education is important only as an investment that will later yield a monetary gain, and this devalues fields of study that will not lead to lucrative jobs in society; it will also devalue the education of minorities and women, who suffer from discrimination in the labour market, since they don't get the same rate of return on their investment in education (cited in Asali Nuseibeh, Forthcoming). Robeyns (2006) also explains that the rights-based approach is also narrow, as it focuses majorly on governments ensuring that the right to education is achieved, and reality shows that many of the countries that are signatories of those rights conventions have vast inequalities in access to education and educational outcomes (cited in Asali Nuseibeh, Forthcoming). Therefore, Robeyns (2006) argues that the "capabilities approach" is a better lens in framing the educational policies.

In the "capability approach", quality education is valuable in itself in enhancing a person's knowledge and way of thinking and learning. Education also enhances other capabilities as it increases people's ability to be healthy, to understand their rights and be able to fight for them, to be able to participate in shaping their environment, in the labour market, in the political arena and on the social level, and most importantly to have an empowered sense of identity (Nussbaum, 2011, cited in Asali Nuseibeh, Forthcoming). Examining educational opportunities from a human capabilities' lens, as Walker (2006: 164) explains, leads us to ask questions such as: are capabilities distributed fairly? Do some people get more opportunities to convert their resources into capabilities than others? Which capabilities matter most in developing agency and autonomy for educational opportunities and life choices? Are all children, young people and adults being taught that they are equally human, or not? How much freedom and opportunities are people endowed with, to choose the kind of life that they value to live? How much agency freedom do people have, and whose achievement and culture is valued and rewarded? And how does education policy and educational services contribute to people's capabilities and well-being?

People in cities need to have the political power to engage in such questions. This is where the "right to the city" concept becomes relevant. The "right to the city" means the ability to participate in decisions affecting urban communities, including the distribution of resources towards educational institutions, and how these institutions function and how inclusive these institutions are, for all the communities residing in that space. The "right to the city" thus seeks to engage members of the community democratically, in the shaping of their educational institutions, and the changing of the culture around education, from one that is directed towards market values solely, to one that is about principals of human flourishing and enhancement of capabilities and social justice. Student bodies, parents' committees, teachers' unions, grassroot organizations, and civil society organizations need to all have a say in the shaping of educational institutions, rather than have decisions lie squarely in the hands of appointed officials, who were not elected by the people residing in that space, or have the decisions squarely in the hands of for-profit institutions. This movement exists already in many cities, and it's composed of countless educators, students, activists and parents who are disillusioned with the neoliberal market experiments in education and unresponsive state control of public institutions (Means, 2014). This movement calls for public schools to be responsive to the diverse needs of children and youth and to be inclusive; it calls for schools that enhance capabilities rather than reduce learning to issues of market competition and standardized test scores (ibid.). Groups fighting for the "right to the city", and the right to urban education for all exist around the world, however with limited freedom of speech in Arab cities these movements are stifled. Let us take a look at the state of education in our four cities.

Education in Arab Cities

Countries of the Middle East and North Africa presented the predominant civilization in education from the eighth to the thirteenth century, and they made innumerable contributions to the world in mathematics, astronomy, medicine, architecture, philosophy and technology (Aubert & Reiffers, 2004). The decline in the region happened as it was unable to compete in a global economy of iron, coal and steam, they also could not follow the industrial revolution and were left behind; scientific thought, based on questioning and systemic experimental research was shifted to Europe, leading to them becoming more advanced economically, technologically and educationally (ibid.).

In an attempt to strengthen the Ottoman Empire (and Ottoman army) and to catch up with western powers, the ruler Mohammad Ali in Egypt sought to modernize education at the beginning of the nineteenth century. He established two educational systems, one for the masses, that attended Islamic schools, and another for elite civil servants (Marlow-Ferguson, 2002). These schools included military training, medicine, pharmacology, veterinary medicine, engineering, arts, irrigation, agriculture, industrial chemistry, gynecology and obstetrics, languages, accountancy and administration (ibid.). Mohammad Ali was aiming to create well-educated, loyal administrators and army officers to serve the Ottoman Empire, and he succeeded in creating a stratified educational system, with different career trajectories and opportunities, one that served the middle class and another that served the lower classes (Hartmann, 2008). In Palestine, during the Ottoman rule, there were disparities in the educational services offered to Muslim, Jewish and Christian children. The education offered to Muslim children was of poor quality and in the Turkish language (Asali Nuseibeh, 2015). As for the Christian and Jewish communities in Palestine, they were granted autonomy to control their own school systems (due to pressure from western powers), and as a result, had superior education compared to the public education offered by the Ottoman administration (Broco & Trad, 2011).

After the collapse of the Ottoman Empire, the European colonial powers spread through the Middle East and Africa, and they decided amongst themselves who would control which territory. Educational policy in the different Arab countries reflected the priorities of the colonizing power. Tunisia, Morroco, Algeria, Lebanon and Syria fell under French colonial control, and the educational policy in these countries reflected France's cultural hegemony, which sought to impose French culture and language, while Palestine, Jordan, Egypt and Iraq fell under British colonial control, which aimed to curb national sentiments and limit educational expansion. The British mandate of Egypt from 1882 continued with the same social and economic stratification policy and a dual education system as that of the Ottoman empire (Marlow-Ferguson, 2002). The colonial power feared unrest by educated Egyptians, so budgets for public education were curbed, curriculums revised and many of the schools that offered free education started charging fees (Hartmann, 2008). The neglect of Egyptian education during the British mandate was so severe that by the end of the mandate in 1922, the majority of the Egyptian population was illiterate (Marlow-Ferguson, 2002). Similar trends were witnessed in Palestine; the British mandate in Palestine inherited a stratified educational system from the Ottoman Empire, with fewer Muslim children attending schools than their Christian and Jewish peers (Broco & Trad, 2011). Although the British mandate changed the language of instruction in Palestinian schools to Arabic, and increased enrollment rates in comparison to the old Ottoman administration, education was neither compulsory nor universal (Tibawi, 1956 cited in Asali Nuseibeh, 2015). Palestinian education today is still fragmented under the Israeli Occupation (Asali Nuseibeh, Forthcoming).

As for Egypt, Jordan and Tunisia, the three countries started rebuilding their education systems after achieving independence. It was Jamal Abdel Nasser who started introducing mass education in Egypt and refining the curriculum, which became a model for the region, greatly influencing other Arab education systems, which often employed Egyptian-trained teachers (Hartmann, 2008). However, although all post revolution leaders aimed to improve the education sector and increase the provision of universal education, the fluctuating political situation in the region and the three wars that Egypt went through followed by recessions and fluctuating oil prices in combination with an exponential population growth meant that resources directed to improve the education sector remained limited (Marlow-Ferguson, 2002); the quality and the relevance of the education offered remains questionable to this day. Although over the last few decades countries such as Egypt, Tunisia, Jordan and Palestine have all sought and succeeded in increasing enrollment rates, there are still major problems, such as educational disparity within the countries and low quality of the education offered (Kamel, 2014). Arab countries are not keeping up with the pace of the rest of the world in education and technology, and cannot hope to do so under present conditions (Aubert & Reiffers, 2004).

Inequality of opportunity in Education in Arab states is high in comparison to other regions (Hashemi & Intini, 2015). Combating educational exclusion requires attention to several issues, among them, the child's physical and emotional well-being, which affect her/his ability to learn, the household and neighborhood environment, and the synchronization of the child's language and culture and that of their educational institution and learning processes (Foster, 1989; McCullum, 1989, cited in Asali Nuseibeh, Forthcoming). Educational disadvantage can be linked to factors inside the institutions of education, and also due to factors outside those institutions, even the maternal diet, health and well-being during pregnancy can have long-lasting effects on the child's health and cognitive well-being (ibid.).

Looking at education through the lens of the "capability approach" in the context of our four cities means looking at the access of the city residents to quality education that enhances their opportunities and their agency freedom to live the life that they have reason to value (Walker, 2006). This means that resources for the education sector need to be distributed to serve all equally and that there is official recognition of the different sociocultural identities within the groups of learners (Tikly & Barrett, 2011), which will require accommodating these differences with an inclusive education system that respects diversity. This requires a broadened kind of inclusive approach that will take into account the gaps that exist as a result of socio-economic status, gender, ethnicity, sects, language, religion and sexual orientation.

Therefore, in assessing inclusive education one needs to start by looking at resource inputs in the different educational settings, such as the availability of school meals in impoverished neighborhoods , complementary supporting programs, psychological counseling and how inclusive and culture and gender sensitive the teaching materials are, including methods and curriculums (Tikly & Barrett, 2011). This requires a nuanced understanding of the different needs of different groups of learners, which is critical for enabling education planners to target resources and interventions effectively (Tikly & Barrett, 2011). It is also important that the residents of these urban spaces have a say in how education in their cities is structured and financed. Education in Arab countries is still a one-size-fits-all. Inequality of educational opportunity can exist at several stages of a child's life. In the following section, we will look at early childhood inequalities in our four cities.

Early Childhood Inequalities

When we talk about early childhood development, we mean a child's development from the time they are in the uterus to the age of five years old (El-Kogali & Krafft, 2015). This period is crucial in determining a person's future health capital, learning outcomes, academic achievements and behavioral and emotional development (Stewart, 2016 cited in Asali Nuseibeh, Forthcoming). Deficiencies in capabilities during childhood not only affect the child's well-being, but can also affect their capabilities as adults (Sen, 1999). Children born to mothers who reside in excluded and impoverished neighborhoods, who do not have access to prenatal care, a healthy diet and a safe environment, can have long-lasting developmental delays. After the child is born, there are biological and psychosocial factors that can affect their development such as cognitive stimulation, iron and iodine deficiency, violence, and maternal depression (Walker et al., 2011). Also, exposure to toxins such as lead or mercury as a result of residing in polluted environments, whether prenatally through the mother, or after being born, through breastfeeding, or direct exposure to polluted water, air and soil, can affect the child's mental development (ibid.). Deficits early in life can be detrimental for a person's well-being and later life trajectories.

Quality early pre-school education can partially remediate the effects of adverse environments and reverse some of the harm that disadvantaged children face; it enhances students' cognitive skills and readiness for primary education by giving them a head-start in language and literacy skills, self-regulation, social-relational competence and early math skills (Heckman & Masterov, 2007; Pianta et al., 2009 cited in Asali Nuseibeh, Forthcoming). Early childhood intervention programs are the most cost-effective educational interventions, because they have a greater impact, at lower cost, than those done later in life (El-Kogali & Krafft, 2015). These interventions are not only beneficial on the individual level, they are also important for the economic and social health of the communities as a whole, as they reduce crime and increase productivity (Pianta et al., 2009 cited in Asali Nuseibeh, Forthcoming).

Pre-school education in the four cities is not mandatory, is very limited in scope and is typically utilized by well-off families. In Egypt, Tunisia Jordan and Palestine, attendance ranges between 20% and 40% of children (Anera, 2020; El-Kogali & Krafft, 2015). In East Jerusalem, the Palestinian Authority does not provide pre-school education and the Israeli Municipality provides free education to a minimum number of Palestinian pre-schoolers, leading to high dependence on private organization to provide educational services to Palestinian pre-schoolers (Asali Nuseibeh, 2015, Forthcoming). This means the majority of pre-schools available are offered to people on a tuition basis, which adds to inequality and the exclusion of the most vulnerable children. The lack of a compulsory and free pre-school education in Arab cities is not only harmful to children, it also delays progress towards achieving gender equality, since women still carry most of the burden of childcare, and if the majority of the available pre-schools are private and charge tuition fees, this will add a financial burden on women wanting to join the labour market (Asali Nuseibeh, 2019). The devaluation of women's work and the gender segregation in the labour market—where women are clustered in low paying jobs—means that for some women its financially more feasible to stay at home and care for children (Asali Nuseibeh, 2019, Forthcoming). The MENA region has low levels of investment in early childhood development, which means that the majority of enrollment—71% in pre-school education—is in private nurseries (El-Kogali & Krafft, 2015).

Jordan faces an additional strain on its already weak pre-education system; the influx of refugees from neighboring countries adds an additional strain on its resources, especially on its health and education systems (El-Kogali & Krafft, 2015). A number of background characteristics relating to the child, such as family wealth, education and geographic location, affect their opportunities to attend pre-school education in the Arab world (El-Kogali & Krafft, 2015). In Jordan, a child in the poorest quintile of the population has an 11% chance of attending pre-school, in comparison to 39% chance for the richest child (ibid.). The poorest Palestinian child has a 24% chance of attending, in comparison to 48% chance for the richest child, and the poorest child in Egypt has a 16% chance of attending pre-school, in comparison to a 65% chance for the richest child (ibid.). The biggest disparity is in Tunisia, where the poorest child has a 13% chance of attending pre-school, and the richest an 82% chance (ibid.). These disparities mean that, depending on the wealth of their households, children reach primary school with different cognitive, emotional and social preparedness (ibid.). In Tunisia and Egypt, children residing in cities are more likely to attend pre-school education. However, in Jordan and Palestine there aren't any significant discrepancies between urban and rural children (ibid.). Another issue affecting early childhood development in our four contexts is violent discipline. Violent discipline of children aged 2–14 is a pervasive problem in the four countries, an issue that hinders child development (ibid.). In the four countries, children are more likely to experience violent discipline, than development activities such as reading, counting, drawing, playing and singing (ibid.). The major reason behind this is the lack of teachers' training and qualifications, to enable them to provide a high-quality development program on the social, emotional and cognitive levels to prepare children for their next educational stage. Pre-school teachers in many cases do not have the education and training needed to work with children. Even for those who do have diplomas or university degrees in education, the quality of their education and training is not of high quality. Education departments at Arab universities have lower entrance requirements, hence attracting people who failed in other fields, and who may have the least capabilities. The devaluation of the teaching profession, and the low salaries offered, also affect the quality of people attracted to this profession, and the reasons behind them resorting to this profession. This can be very damaging to cohorts of students, since early childhood education, health and cognitive development greatly affect the student's chances in school later. Let's take a look at the available primary and secondary school experience of children in our four contexts.

Educational Inequality at the Primary and Secondary Levels

Unterhalter (2003) brings to our attention that education in Amartya Sens "capability approach" is undertheorized, as it does not take into account the unequal social relations within schools, and the different processes of teaching, learning and management within school institutions, that can have varying consequences and outcomes on various groups within the institution. She argues that the school experience, in certain contexts, can be a cause of capability deprivation, rather than of human capability development. Therefore, one needs to look beyond school enrollment rates, to look at the quality of the school experience of the students, and how much it builds their capabilities and enhances their agency freedoms.

The four cities have similar educational systems. Students start primary education at the age of six. In both Egypt and Tunisia, the primary stage lasts six years and is followed by a three-year preparatory phase. Primary and preparatory schooling comprise compulsory basic education, and after that the students either track into vocational secondary or general secondary (Krafft et al., 2019). In Jordan and Palestine, the basic education stage lasts ten years and is followed by two-year secondary education, either in the vocational track or in general secondary, which is the academic track leading to higher education. In

Egypt, general secondary essentially guarantees access to higher education, while in Jordan, Palestine and Tunisia examinations at the end of general secondary determine access to higher education.

Educational enrollment in all four countries is high, with almost universal enrollment in Palestine and Tunisia, followed by Jordan with around 90% enrollment, and then Egypt at 75% enrollment (Assaad et al., 2019). Amongst Palestine, Jordan, Tunis and Egypt, the difference in the probability of entering school between rich and poor children is highest in Egypt, meaning it has the greatest inequality of opportunity in initial entry to school (ibid.). As for equality of opportunity in entrance to secondary education, the gap between the rich and poor is highest in Jordan, followed by Tunisia, then Egypt and Palestine (ibid.).

> I went to a school in a poor area and the school was out of control, meaning nobody was following any system and the teachers were not that good, and they did not help that much. It is a public school in the end. If you ask anyone who went to a public school they will tell you the same....When you pay the price, people will want to look after you more. In public-schools people take the government salaries, which are less than my salary now. The teacher will not give his all.[2]

In the four countries, there are also issues with the quality of the education offered. Hashemi and Intini (2015) studied education quality in eleven Arab countries—among them Tunisia, Palestine, Egypt and Jordan; through the math and science test scores of fourth-grade and eighth-grade students, reported by the Trends in Mathematics and Science Study (TIMSS), in 1999, 2003, 2007 and 2011, and test scores in reading, reported by the Programme for International Student Assessment (PISA), in 2000, 2003, 2006, 2009 and 2012. The authors revealed that overall, performance of Arab countries in all subjects is lower than in non-Arab developing countries. Added to that, there is persistent inequality when it comes to parents' educational level and socio-economic status. Their results show a strong and significant relationship between the parents' education and the child's educational outcome.

[2] Interviewee Mahmoud (Pseudonym), 25, Muslim, Male, Marketing Employee, Amman.

> The activities that build our skills outside school are non-existent, especially if the parents are not educated, they don't know how important it is to build these skills for the child.[3]

Children in more educated households performed significantly better than those with illiterate or less-educated parents, particularly for grade eight. The study also showed that children who had more books at home significantly outperformed children in households having few or no books at home. In addition, children in affluent families performed better in math and science than children in disadvantaged families. Salehi-Isfahani et al. (2014) did a similar study exploring inequality of opportunity in the education sector in sixteen MENA countries, by the use of the TIMSS exams held in in 1999, 2003 and 2007. They discovered that family background is the most important determinant of education opportunities, followed by the region of residence.

Policies of privatization that swept through the Arab world, and cuts to social programs, among them public education, have exacerbated the inequality problem. In Egypt, the majority of students—over 90%—attend public (governmentally funded education) and the rest attend either Islamic religious Azhari schools, or private schools (Krafft et al. 2019). Although Egypt has achieved great success in improving enrollment rates over the years, the problem remains with the quality of education, and the relevance of the education Egyptian children receive. Parents are expected to supplement free governmental education with private tutoring, because the quality of public education is low. This places a financial burden on families looking to enhance their children's chances of academic success (Assaad & Krafft, 2015). There are two main types of private tutoring; official after-school classes, provided by the schools' teachers after school time, and organized by the Ministry of Education, at a lower price. The second type of tutoring comprises of private classes offered at a higher price at the students' houses or in established tutoring centers (Sieverding et al., 2017). In Cairo, in particular, the widespread reliance on private tutoring sheds doubt over the availability of free public education (Assaad & Krafft, 2015). For Cairenes children, for example, at the preparatory level, 35% of Egyptian students took private lessons on the national level, compared to 59% in Cairo (Sieverding et al., 2017). Likewise, while nationally 68%

[3] Interviewee Lima (Pseudonym), 25, Muslim, Female, NGO Employee, Tunis.

of general secondary students took private lessons, 77% did so in Cairo (Sieverding et al., 2017: 573). Receiving supplementary private tutoring is linked to socio-economic advantage. For example, children of parents who have acquired higher education, and with higher income, were more likely to receive private lessons (Assaad & Krafft, 2015). The main drivers behind the prevalence of private tutoring has been associated with the low quality of the public school education system in Egypt. Public education in Egypt has expanded exponentially from serving 1.9 million students in the 1950s to serving 23 million students today (Sieverding et al., 2017). This increase in demand was not met with a similar increase in budgets, which meant that the system was not prepared to cope with the growth in the student population, leading to overcrowding, and low quality of educational services. Sobhy (2012) argues that the political and economic climate in Egypt during the Mubarak Era, that pushed for privatization, exacerbated the problem. Mubarak's policies of reducing public spending on education has increased the growth of private tutoring, where underpaid teachers are disincentivized to offer quality education, and try to supplement their low income by offering private lessons, adding pressure on poor families, which constitute about 40% of the population (ibid.). In some cases, school administrators took an active part in promoting private tutoring, by presenting private classes as obligatory to students (Sieverding et al., 2017). Private tutoring can be an exhausting financial burden on poor families, who are estimated to spend a fifth of their income on schooling, which is meant to be offered for free (Sobhy, 2012).

This privatization of education has undermined the public education system severely; teachers take private tutoring more seriously, while students do not take the official school day seriously, resulting in high levels of absenteeism, especially at the secondary level (Sieverding et al., 2017). Mubarak's policy of privatizing education, and spending less money on pre-university education than on the university education, meant that many students of low socio-economic backgrounds did not get enough financial support to access higher education (Amira, 2017). Elite universities require high grades in the secondary exam, so the parents' ability to spend on their students' primary and secondary education, whether through sending them to private schools, or spending money on private tutoring, determined the students' ability to access elite publicly funded universities (Amira, 2017), meaning the rich are benefiting more from governmental funds spent on higher education. The school education system will remain weak, as long as the government keeps the current teachers' salary scales. The World Bank recommends that an average teacher salary of 3–3.5 times the GDP per capita is conducive for a productive education system; in Egypt the average teacher salary is 1.3 times the GDP per capita (Johnson, 2018). If teachers' salaries remain low, teachers will keep depending on private tutoring. Egypt's failure to invest more in school education comes as a result of a combination of political and economic policies, and partially its extensive borrowing. For the fiscal year 2018–2019, interest payments on foreign loans alone encompassed nearly 55% of all revenues, eroding social expenditure (ibid.). In 2020, Egypt has managed to secure financing agreements worth $252 million, from several donors in an aim to digitize education by providing servers and screens and digital teachers training, at public schools (Karima, 2020). However, as long as major problems in the Egyptian educational system persist, these programs remain superficial. The major infrastructural problems, overcrowding, teachers' qualifications and salaries, are all major and highly expensive issues that need to be addressed, in order to make a difference in the quality of education provided to Egyptian children.

Tunisia has a similar story of shifting towards privatization of the education sector. Following its independence from French colonization, Tunisia has achieved great progress in developing its education sector. Enrollment rates went from 14% in 1958 (Milovanovitch, 2014) to almost universal enrollment today. Bourguiba's Era witnessed great investments in the education sector in Tunisia, from 14% during the French colonization in the 1950s to 36% of the Bourguiba's government budget in 1970s (Masri, 2020). Educational development went hand in hand with economic growth, and the country managed to increase enrollment in primary and secondary education, and expanded the institutions of higher education. Bourguiba also managed to strengthen vocational training, which was an important factor in increasing employment levels in Tunisia. He also made co-education the norm in Tunisia, and set an agenda to achieve gender equality in education (Masri, 2020).

The seeds of the Tunisian ambitious postcolonial education agenda can be traced back to 1840, with the establishment of the Ottoman Bardo Military Academy, which operated independently from religious authority, and encouraged critical thinking skills and the tradition of debating (Masri, 2020). In 1875 Sadiqi college was established, which was one of the most prestigious learning institutions, as it sought to preserve Arab culture, while emphasizing sciences. The college accepted students only on merit-basis and had students from all backgrounds, and a third of its student-body were Jewish Tunisians (ibid.). Educational development continued with the establishment of Al-Jami'iyya Al-Khalduniyya, and the Association des Anciens Élèves du College Sadiqi, which were dedicated to teaching sciences (ibid.). When Bourguiba took charge, the seeds of a strong education system were already planted, and he continued with a policy that prioritized education.

However, the combination of neoliberal policies, and two decades under the dictatorship of Ben Ali, eroded the education sector in Tunisia, leading to low-quality of learning outcomes, and a lack of public conviction in the institutions of education. In the year 2012 alone, after the ousting of Ben Ali, there were 300 corruption allegations at regional and national levels, as well as in schools (Milovanovitch, 2014). Ben Ali's privatization policies of the education sector resulted in cuts to public funding to education, which led to the flourishing of private tutoring and 70% of 15-year-olds in the country receiving private tutoring (ibid.). "If you don't get private tutoring in your final year, you are literally crawling and struggling to pass your final exams".[4]

An increase in enrollment in private schools was also witnessed. Similar to the Egyptian problem with private tutoring, when it becomes so prevalent and a necessity for success in exams, teachers start expecting students to take private tutoring. This demotivates both students and teachers from putting an effort in the regular school day. The public education system starts to have an under the table parallel private system, adding financial burdens on parents and creating gaps between students with different socio-economic statuses. These issues erode educational equality and increase the gaps in later life trajectories, such as the ability to enroll in institutions of higher education, the ability to secure employment, residential options and overall well-being.

> They say at the university level that it's all equal. How is it all equal? when the skills I acquired from school are different.... You sit at the university hall, and you see how more advanced private school students are. The private school children, their English and French are better than ours. They start learning languages much younger than us.[5]

[4] Interviewee Sina (Pseudonym), 23, Muslim, Female, Dental Lab Technician, Tunis.

[5] Interviewee Lima (Pseudonym), 25, Muslim, Female, NGO Employee, Tunis.

Jordan has similar trends in the education sector. However, it has a refugee crisis, which adds pressure on its already strained resources. In 1921, when the Emirate of Transjordan was established, educational facilities inherited from the Ottoman Era consisted of religious schools that provided limited education. Although education started to expand in the following two decades, the budgets offered to education by the British colonial power were very limited. After its independence, Jordan sought to expand its education system and reduce prevailing illiteracy rates. In 1952, Jordan made primary education compulsory and free. Jordan, similar to Tunisia and Egypt, managed to achieve almost universal school enrollment at the primary level. However, inequality still persists, with disabled children, refugee children and children from lower socio-economic statuses being disadvantaged, and are at a higher risk of being out of school. Enrollment at the secondary level in Jordan is lower than that at the basic level (Grades 1–10). At the secondary level, girls are more likely to be enrolled in schools than boys, despite the fact that some girls are forced into early marriages or staying at home to care for younger siblings or ailing family members (ibid.). Poverty is also a major driver for early marriages, a study by UNICEF shows that early marriage has been used as a coping mechanism for Syrian refugees in Jordan, with one out of three newly registered marriages of Syrians in Jordan involving a child, under the age of eighteen (UNICEF, 2020). Being a boy means a 7% increase in the risk of dropping out of school in Jordan (ibid.). Boys are more likely to drop out of school as a result of push and pull factors. Push factors manifest in violence and bullying at schools, and a lack of any meaningful academic experience, that result from poor resources and unqualified teachers.

> My school and most schools I find are too traditional and they don't provide us with the needed skills to continue with our lives after... Teachers have memorized the knowledge and they don't understand it. With all due respect, they are also the product of what came before them.[6]

Pull factors manifest in the demand for cheap and unskilled labour, as young boys from lower socio-economic statuses are expected to start supporting their families at younger ages. There are many challenges in

[6] Interviewee Mahmoud (Pseudonym), 25, Muslim, Male, Marketing Employee, Amman.

school education services offered to Jordanian children, amongst them overcrowding, weak infrastructure and underqualified human resources. There's a difference between governmental schools and private schools, with private schools having better school facilities and infrastructure. The majority of children who are enrolled in schools in Jordan attend governmental public schools 68.2% (UNICEF, 2020). Almost a third of students, 26%, attend private schools, and the remaining 5.8% attend the United Nations Relief and Works Agency (UNRWA) schools, which are schools serving Palestinian refugee children (ibid.). The inflow of 225,000 Syrian students since 2011, added a strain on an already overcrowded education system (Stave et al., 2017). This has led to an increase in the number of double-shift schools, which accommodate Jordanian children in the morning and Syrian children in the evening (ibid.). Overcrowding in schools not only results from the influx of refugee children, it is also due to the uneven distribution of schools in Jordan, with 11% of students attending 43% of schools, and 89% of students attending 57% of schools, mostly located in urban localities (ibid.). Overcrowding leads to the degradation of school property and facilities. It also affects the overall student experience and the teachers' ability to attend to students' different needs and abilities, leading to low learning outcomes. Students at the primary level are not performing well, with 80% of second and third graders reading without comprehending (National Committee for Human Resources Development, 2016 cited in UNICEF, 2020). Also, 15-year-old's performance in the PISA exam in 2018 was lower than the average, in reading literacy, mathematics and science (OECD, 2019), ranking them among the bottom ten countries. Jordanian students' achievement in the TIMSS exam also declined between 2011 and 2015, by 20 points in mathematics and 23 points in science (Mullis et al., 2016). Apart from scores in standardized tests, the quality of the educational experiences offered in Arab countries is ranked among the lowest in comparison to other countries. Quality education measures, such as the World Economic Forum's Global Competitiveness Report, examine the extent to which educational institutions equip their students with the ability to think critically and creatively, and how effectively these institutions foster students' curiosity (World Economic Forum, 2018) looking at our four contexts, they do not score well in the report, with Egypt scoring the least among them. The World Economic Forum's Global Competitiveness Report for 2017–2018 ranked Egypt's quality of primary education as 133rd out of 137 countries measured.

> If I have to evaluate the education system in Egypt, I will say that, I see it as a failed system unfortunately. It's all based on memorization. When I compare between me and my friends who studied at international systems, I feel that there are so many things, as a human being, I should know and acquire from the education system, but I didn't.[7]

The World Economic Forum's Global Competitiveness Report for 2017–2018 ranked Tunisia 83 out of 137, and Jordan ranked 60 out of 137. They did not provide data on Palestine. Palestinian education is a unique case as it is still functioning under occupation.

Throughout the last five hundred years,[8] Palestinian education has been perpetually underfunded and only catered to serve the ruling bodies' interests; it was not aimed at improving political, social and economic capabilities of Palestinians. During the Ottoman rule, Palestinians received poor quality education in the Turkish language, disregarding their Arabic mother-tongue (Asali Nuseibeh, 2015). In the nineteenth century, as the Ottomans were starting to lose power, western powers started to intervene in their internal affairs, especially in matters relating to non-Muslim minority populations (Khuluq, 2005). Christian and Jewish communities in Palestine started to gain control over their own school systems; churches were able to establish modern schools, superior to public schools offered by the Ottoman administration, and they prepared the students for entry into western colleges (Broco & Trad, 2011). Elite Muslim families were able to enroll their children in the private church schools. This led to educational inequality along religious lines and socio-economic lines. There was also educational inequality along gender lines, and limited provision of educational services overall, which meant that by the end of the Ottoman era, less than a quarter of Palestinian school-age children (17,000 [of which 13,000 boys and 4000 girls] out of 73,000) attended school, with the majority attending private schools (Demichelis, 2015: 266).

Even with the dire educational climate, elite Palestinian intellectuals, both Muslims and Christians, took great part in the cultural awakening or

[7] Interviewee Maha (Pseudonym), 18, Muslim, Female, Political Science Student, Cairo.

[8] This section is taken verbatim from: Asali Nuseibeh, R. (Forthcoming). Education in Palestine. In Badran, A. and Dumper, M. (Eds.), (Forthcoming) *The Routledge Handbook on Palestine*. (1st ed. London, 2023: Routledge.)

renaissance movement called *An Nahda*, which took place in all Ottoman-ruled Arabic-speaking regions, between 1870 and 1950. This period of intellectual revival, inspired new models of political resistance, calling for the end of colonialism, social reform and gender equality (Pormann, 2006). The Governmental Arab College of Jerusalem,[9] a school established in 1918 during the British mandate, played a role in this intellectual revival, hosting several prominent Palestinian and Arab literary figures. The school had several prominent principles, some of which had a contentious relationship with the British government. For example Khalil Al Sakakini's tenure as the director of the school was short lived, because he quit in protest against the British appointment of Herbert Samuel as High Commissioner of Palestine (Demichelis, 2015). Palestinian intellectuals were continuously frustrated by British support to the Zionist project and the underfunding and censorship of Palestinian education. Thirty years after the establishment of the college, in 1948, it was closed, and many of its students became refugees.

Although this school was considered one of the leading educational institutions in Palestine, Palestinian education during the mandate was neglected and underfunded. The British Occupation of Palestine started in 1917, but it wasn't until 1920 that the Civil Administration (the organization responsible for delivering services to the public) began to work on the education sector (Broco & Trad, 2011). Although the British Occupation changed the language of instruction in Palestinian schools to Arabic, and increased enrollment rates in comparison with the previous Ottoman administration, education was neither compulsory nor universal; also gender, religious and urban–rural inequality persisted (Tibawi, 1956). During the British mandate there were two education systems, one for Jewish people and one for Palestinians, with the Jewish community having full autonomy and international support for their education, while Palestinian education was completely controlled by the British Mandate (Broco & Trad, 2011). While Jewish education was given high fiscal priority, and the curriculum was developed to serve nationalistic aims, Palestinian education was given lower budgets, and set minimum standards for school attendance and was apolitical to curb nationalistic views and uprisings (ibid.). The British policy in Mandate Palestine was to hinder any state-building efforts by Palestinians, which

[9] It was first referred to as Men's Teacher Training College, until 1927 when the name was changed to Governmental Arab College (Demichelis, 2015).

could potentially compete with the Zionist project (Brownson, 2014). For example, the British Mandate adopted a policy of classifying the population by religion and not nationality; Muslim, Christian, Jewish and others such as Druze. Furthermore, Palestinian educators were unable to obtain from British authorities the permission to establish an Arab University of Palestine, while as early as 1925, Jewish educators were able to obtain the British Mandate's support to establish the Hebrew University of Jerusalem (Demichelis, 2015).

British officials wanted to prevent revolts by educated Palestinians, based on previous experience in areas they colonized, such as Egypt and India, where the educated elite revolted against them (ibid.). The deliberate British policy of neglect and censorship of Palestinian education was a constant source of frustration to Palestinian intellectuals, and by the end of the mandate a mere third of the school-age population (93,550 out of 320,000) was registered in elementary and secondary institutes (Demichelis, 2015: 266). Due also to the UK's policy of population classification by religion, many foreign Anglican and Protestant Schools were established, also, various Christian denominations supervised the education of their children, Latin, Orthodox, Syrian and Armenian Patriarchs, the Custode di Terra Santa and the Archbishop of the Greek Catholic Church (Broco & Trad, 2011: 5). The Jewish community of Palestine at the time had a separate Hebrew educational system as well, which was affiliated with the international Zionist movement which financed them, in three trends, secular-socialist, religious-Zionist and non-Zionists ultra-orthodox (ibid.). This meant that educational inequality persisted during the mandate, and by the end of the mandate in 1948, Jews and Arab Christians were completely literate, while only 40% of Arab Muslims were (Demichelis, 2015).

After the 1948 war, Jordan took charge of education in East Jerusalem and the West Bank, and Egypt took charge of education in Gaza for non-refugee children. They offered public education from grades one to twelve and instituted the *Tawjihi* matriculation examination. As for Palestinian refugees residing in East Jerusalem, West Bank and Gaza they received education at schools supervised by the United Nations Relief and Works Agency (UNRWA), from grades one to grade nine. Private schools supervised by charities and religious institutions remained serving students on a tuition fee basis.

After the 1967 war, Israel took over the functions of the education ministries of Egypt and Jordan in what is known today as the

Occupied Palestinian Territories (East Jerusalem, the West Bank and Gaza). Although the Egyptian and Jordanian curriculums were maintained, Israeli policy was to suppress and censor teaching about Palestinian culture and history (Demichelis, 2015). Israel, as an occupying power, was not interested in developing Palestinian education. The combination of a weak education system and an economy stifled by the occupation drove Palestinians to joining the Israeli labor market as unskilled or semi-skilled labourers. It also pushed for early marriages, which played a role in pushing the Palestinian society towards conservativism (Broco & Trad, 2011). The first intifada of 1987–1993 exacerbated the problem in the education sector, where Palestinian schools were destroyed, and many were closed for extended periods of time, teachers and students were detained, and many lost their lives.

Following three decades of Israeli military control, the PA inherited a weak educational system. Since its establishment in 1994, the PA exerted efforts to improve the provision of educational services to the Palestinian population, by providing over a thousand new schools. It also established a Palestinian curriculum and expanded the teaching cadres, and worked on increasing the students' gross enrollment and completion rates. It also worked on reducing overcrowding and eradicating the two and three shift school systems. However, the PA is working in a limited capacity, as it has limited authority and sovereignty over its land,[10] and is highly dependent on donor funding, hence, the Palestinian education sector today remains fragmented and underfunded. The internal Palestinian strife between Fatah and Hamas, and the siege that Israel imposed over Gaza, resulted in the establishment of two parallel educational administrations, one in Ramallah and one in Gaza. At the national level, the Ministry of Education and Higher Education (MoEHE) is based in Ramallah and is responsible for overseeing the education sector across the West Bank through its 17 district offices, it also monitors the provision of education by private schools (UNICEF, 2018). There is also a parallel MoEHE office in Gaza City that supervises the provision of

[10] "The West bank—where around 61 per cent of the Palestinian population lives—is divided administratively into three areas in accordance with the Oslo II Accord. Area A is exclusively administered by the Palestinian Authority; Area B is under Palestinian civil control and joint Israeli-Palestinian security control; Area C is under Israeli security and administrative control, except for education and health services which are provided by the Palestinian government" (UNICEF, 2018: 9).

education services in public schools in Gaza through its 7 district offices (ibid.). As for the education of Palestinian Israeli citizens inside Israel's 1948 borders, it is governed by Israel, and is a separate system that teaches an Arab–Israeli curriculum. The education of Palestinian refugees is governed by UNRWA and supervised by the hosting countries. As for the education of Palestinian residents of Occupied East Jerusalem, it is the most fragmented system, with several bodies controlling the education of over a hundred thousand Palestinian students residing in the city. Israel prohibits any Palestinian Authority activity in the city and so the Palestinian Authority controls thirty-eight schools indirectly through the Awqaf system. The number of students in those schools are dwindling every year as they are the most deprived and least funded schools in the city. Although the Palestinian Authority has managed to expand the education sector in the West Bank and Gaza, its educational institutions in East Jerusalem have shrunk over the years. School education in East Jerusalem has its own set of problems; it is fragmented, with different supervising bodies and there is a lack of coordination or common vision amongst them. There are 38 schools governed by the Palestinian Authority, 6 UNRWA schools, 64 private schools and 74 Israeli municipal schools (PCBS, 2017). Even though there are several providers of educational services, there is a lack of 2,557 classrooms in East Jerusalem (Tatarsky & Maimon, 2017), and drop-out rates are high, reaching 1,300 students annually, with the majority dropping out when they transition to 10th grade and over (ibid.). The Israeli municipality of Jerusalem claims that the lack of available educational facilities in East Jerusalem is the result of a lack of land. However, this scarcity is a result of the discriminatory policy that has been implemented by the Israeli Jerusalem municipality since 1967, which has confiscated over 38% of Palestinian East Jerusalem for the construction of settlements that serve Israeli citizens, in addition to refraining from planning any expansions of Palestinian neighborhoods the way it expands and develops Israeli neighborhoods (Tatarsky & Maimon, 2017). Since the beginning of 2009, under the Israeli Prime Minister Benjamin Netanyahu and the Jerusalem Mayor Nir Barkat, 10,000 housing units have been approved for Israeli settlements in East Jerusalem, in comparison, no broader outline plans have been approved for Palestinian neighborhoods, with only minor housing units approved (Tatarsky and Cohen-Bar, 2017). Lack of planning of Palestinian neighborhoods has resulted in only 2.6% of all the land in East Jerusalem being designated for public buildings and only 15% of the area of East Jerusalem is zoned for Palestinian residential building, which constitutes 8.5% of the area of Jerusalem as a whole (ibid.). This inadequate planning and inadequate budgeting of Palestinian neighborhoods prevents the construction of new schools; for example almost 50% of students in A-Tur neighborhood cannot find places in city schools (ibid.).

Governance of Urban Education

The combination of the "capabilities approach", and the "right to the city", calls us to not only look at how the education system is building the students capabilities, but to also look into the governance of the education sector in our four contexts. It pushes us to ask questions on how the education sector is managed, to what extent do the people in the four cities have the political power to engage with questions related to the distribution of resources towards educational institutions in their cities, how these institutions are functioning, and how inclusive these institutions are for all the communities residing in that space. Is it a democratic process where student bodies, parents' committees, teachers' unions, grassroot organizations and civil society organizations have a say in the shaping of educational institutions?

The reality on the ground shows that education in Tunis, Cairo and Amman is still highly centralized with a top–down approach to governance. In East Jerusalem, almost fifty percent of school education is provided by private actors, so it is less centralized than the other three contexts, but it is not necessarily democratic nor equal.

Over the last several decades, there has been a clear global trend towards the decentralization of education, which is a governance strategy of giving local or regional authorities more autonomy in regards to educational policy and the use of resources. There isn't a universal decentralization scheme that countries adopt, rather there are various configurations of decentralization schemes (Channa, 2015). In practice, there isn't an education system that is completely centralized or decentralized, variations exist in terms of the division of responsibilities between central and local authorities (Leer, 2016). Types of decentralizations include delegation, de-concentration, devolution and privatization (Rondinelli et al., 1983). These different schemes range from partial transfer of authority, to complete transfer of authority to local actors.

Decentralization strategies vary depending on the socio-political environment, cultural dynamics, historical context, the administrative structures within each country and the local resources available (Simão et al. 2016). There are prerequisites that are required for education decentralization to result in a positive impact, such as local governments having the political will, the resources and the institutional capacities to respond to local needs, as well as the central government having the political will and the capacity to support local actors in shifting powers and authority to them (ibid.).

Decentralizing education governance has been a hotly debated issue around the globe. Researchers who are pro decentralization have argued that it could reduce operating costs, increase innovation, reduce bureaucratic inefficiencies, empower local authorities, increase accountability and transparency and potentially improve educational output (Simão et al., 2016). Pro-decentralization arguments claim that local authorities are more in touch with the local context, and less restrained by the central state bureaucracy, so they are better equipped to provide higher quality educational services and ensure that schools are more aligned with local educational needs and preferences (ibid.). Decentralization also has the potential to empower local communities, by giving them greater voice, and by including less-advantaged groups in holding schools accountable for providing quality education that is relevant for all (ibid.). However, evidence from various studies on the effects of decentralization are context specific and cannot be applied to all countries equally (Kameshwara et al., 2020).

Arguments against the decentralization of the education system, center around the fact that decentralization is part of the neoliberal reforms that have swept the globe. The assumption is that by allowing schools to operate in a quasi-market framework by reducing the government role, the schools will be more flexible, innovative, and responsive to local needs (Leer, 2016). Arguments against the decentralization of the education system include the potential of it resulting in increasing local divisions, strengthening elite power, opening the space for corruption and nepotism, and reducing economies of scale. It could also lead to increased inconsistencies in curricular and quality standards, and result in the fragmentation of the education system, as well as reduce compliance with national policy initiatives (Simão et al., 2016). Hanushek et al. (2013) have studied datasets of PISA tests, spanning 2000–2009, comprising over one million students in 42 countries, and they suggest that autonomy may be conducive to student achievement in well-developed systems, and in countries with strong education sectors, but detrimental in low-performing systems in countries with weaker education systems.

The three decentralization models that have been promoted in Egypt include community schools, schools-based management and public–private partnership (Allam, 2021). Community schools are established and managed by local communities and non-governmental organizations to serve marginalized groups (ibid.). The second model is school-based management, which involves devolving authorities to school management committees that include the school principal, teachers, parents and other community members (ibid.). The third model, public–private schools partnership, involves the transfer of some authority over school management to the private sector and non-state actors, such as philanthropic and faith-based organizations (ibid.).

Decentralization models were introduced in Egypt by international organizations. In the 1980s', the World Bank provided structural adjustment loans to developing countries, among them Egypt, on the condition that they follow its policies of reducing the role of central governments and decentralization, channeling funding to primary education based on an economic model of rate of return on investment, and increasing the private sector's role (ibid.). In the decades after, the World Bank, UN organizations, USAID and NGO's have experimented with different forms of decentralization in the education system in Egypt. Allam (2021) explains how international organizations used coercion through funding, and pressure to overcome resistance within the Egyptian bureaucratic systems. She explains how external interventions have caused a persistence of policies that did not work well in the Egyptian context. The decentralization models adopted were never fully implemented, and there was no significant effect on students' learning outcomes or quality of education. Interventions that worked in western contexts are not necessarily suitable in Arab cities. Allam (2021) advises that Egyptian officials should develop a more critical stance towards foreign experts' interventions, and should listen more to local actors and grassroot organizations.

Jordan faces similar pressure from international organizations to shift its education policy towards a decentralized model. Karmel et al. (2016) say that there's a risk in Jordan that education could be decentralized, not with a focus on educational outcomes, but as a half measure that gives lip service to international support for decentralization. For decentralization to work effectively in the education sector, local authorities and schools need to have the capacity and the resources to be able to handle planning, and to implement educational programs delegated from the central government. The development of legislations, together with political will, enabled both Egypt and Jordan to take the first steps towards decentralizing their education systems. However, the more complex aspects are the capacity building of local authorities and schools, as well as the provision of resources and mechanisms for local authorities to generate resources, and improve the quality of education. In 2010, in its drive towards decentralization, Jordan started with capacity-building trainings in schools across Jordan. Despite the fact that there were attempts to enhance management capacities of schools, the efficiency and quality of schools have not significantly improved, and schools were not given the requisite administrative and fiscal authority (Karmel et al. 2016). With neither the resources, nor decision-making powers, principals and teachers are limited in their ability to improve their schools. Even with the several decentralization initiatives taking place in Jordan and Egypt to democratize educational governance, education remains centralized.

The overall education structure in Tunisia reveals a highly centralized structure as well, strongly controlled by the federal government. However, there has been a shift towards decentralization in the last few decades to shift powers to regional administrations. The organization of a "school project" (projet d'ecole) is one of the decentralization attempts taking place in Tunisia (Akkari, 2005). This project came as a result of an education policy to give Tunisian schools the opportunity to define their own specific projects, based on the specific particularities of each school environment, while involving all stakeholders (ibid). This meant that each school had a status of an independent educational unit accountable for its own results and outcomes, with students, parents, teachers, directors and inspectors reporting on the school performance. These initiatives can be beneficial in improving the learning experience of students in schools, but they remain small in scope in a still highly centralized system. Escardíbul and Helmy (2015) analyzed the effects of decentralization and school autonomy on the quality of education in two MENA countries, Jordan and Tunisia. They found that autonomy management has had no significant effect on student attainment in both countries, except for a minor negative impact in Jordan. Akkary (2014), who looked into education reform in Arab countries, recommends that there should be a paradigm shift in educational reform, meaning breaking the old patterns of uncritical adaptation of western practices and ideas, and engaging local actors and scholars, policymakers and school practitioners, in designing reform initiatives that are grounded in the specific problems and cultural contexts of their schools. So far, the top-down approach to reform has resulted in a culture where reform is viewed as the sole responsibility of national governments, with teachers becoming passive, viewing reform as something happening to them rather than initiated by them (Bashshur, 2005 cited in Akkary, 2014). This learned passivity amongst teachers is aggravated by the belief that taking initiative might upset people in higher positions, causing them to face retaliation, which has turned teachers into bureaucrats, implementing government policies, rather than leaders in their institutions, initiating change (ibid.), hence rendering any decentralization schemes shallow. This problem is not limited to Arab countries only, Akkary (2014) explains that in most postcolonial countries, local knowledge is rarely recognized as being of value, and is highly disregarded, that is why policymakers rely on the knowledge base that is mostly imported, and highly disconnected from the local realities.

One of the major problems in the education system in Arab countries, is its human resource pool. Teachers, principals, administrators and social workers are educated and trained in the same weak system that they are working in. Teachers lack pre-service training and some do not even have a degree in education.

> The teachers, they have a lot of issues. But they didn't get a proper education themselves, and they weren't raised properly themselves, and they didn't get exposed.....and so they don't know other than that, just how to get the worst out of the students.[11]

Teachers are also burdened with overcrowded classrooms, weak infrastructure and high percentages of children coming from impoverished neighborhoods needing extra support and individual care. As we have discussed before, the majority of children also come with no pre-school education and preparation, so they need extra support from teachers, who are not well trained. Added to that, teachers are expected to handle reform initiatives that are imposed on them by higher up officials. Teachers' capabilities are not built to allow them to engage in inquiry, critical

[11] Interviewee Salim (Pseudonym), 19, Muslim, Male, Political Science Student, Cairo.

thinking, or using technology and generating innovative ideas to initiate improvement in their schools (Akkary, 2014). So as long as teachers, principals and local education actors lack the capacity and the resources to implement change in the schools, initiatives for decentralization will remain superficial and, in some cases, even counterproductive.

In East Jerusalem, around half the provision of school education is by private institutions. The governance of education in East Jerusalem is fragmented, with different supervising bodies including the Israeli Municipality, the Palestinian Authority UNRWA and private and religious institutions and charities (Asali Nuseibeh, 2015). The fragmented political governance between the various governing bodies has caused a major issue in the provision of school education, with a lack of available classrooms and overcrowding in the existing ones (see Asali Nuseibeh, 2015). Autonomy of the education sector is a major issue for Palestinians. Palestinians residing in Israel's 1948 borders lack any meaningful autonomy over their education. Palestinians living under the authority of the Palestinian government have educational autonomy; however, the Palestinian curricula are constantly reviewed by western governments and international organizations to censor for any anti-Israeli sentiments, the same level of scrutiny is not given to Israeli curriculums (Asali Nuseibeh, 2015). What is being taught in East Jerusalem schools is a mixture of a censored Palestinian curriculum by the Israeli authorities in Israeli municipal schools serving Palestinian students, and an uncensored Palestinian curriculum being taught in private schools, Awqaf schools and UNRWA schools in East Jerusalem. The fact that half the school provisions are provided by private actors and not the government has not yielded positive results on the education sector in East Jerusalem. Private schools are selective, they have entrance exams screening out children with disabilities and they charge tuition fees, screening out children from lower socio-economic statuses. Schools that are affiliated with churches also give preferential treatment to the Christian minority of Jerusalem, who in return face discrimination when joining public schools. The fact that there are several supervising authorities, with half the education provision privatized means that education remains far from democratic and highly unequal in the city.

Conclusion

This chapter started with a utopian vision of an education system in the city that aims at building people's capabilities, enhancing individual empowerment and achieving sustainable economic development, while increasing community engagement and active citizenship. It calls for a paradigm shift in the view of educational policy from focusing solely on economic returns of education to look at education through the lens of the "capability approach"; to see to what extent education builds people's capabilities and builds an empowered sense of agency. It also calls for an education system that serves all city residents, regardless of social class, ethnicity, gender etc. To create this kind of education system residents of the city need to be able to participate in decisions affecting their lives in the urban space, including the distribution of resources towards educational institutions, and how these institutions function and how inclusive these institutions are for all the communities residing in that space. The concept of the "right to the city" was then presented which seeks to engage members of the community democratically in the shaping of their educational institutions, and the changing of the culture towards principals of human flourishing and enhancement of capabilities and social justice.

In our four contexts, there have been great achievements in terms of enrollment in education. However, there are several problems in the education sector with regards to pre-primary education. Enrollment in pre-primary education remains unequal and highly dependent on the socio-economic status of the parents. Provisions in pre-primary education are still very much provided by the private sector, hence are mostly available to children of families from higher socio-economic status. As for basic and secondary education, although it is provided mainly through the government in Cairo, Amman and Tunis, it is still unequal, as there is a parallel system of private tutoring leading to inequality of opportunity. As for East Jerusalem, around half of school education in the city is provided by private actors, also leading to inequality. The quality of education provided in the four contexts is not of high quality. There are problems in infrastructure, teachers' qualifications, educational programs that do not arouse students' curiosities, and that generate a lack of understanding of opposing views and ability to critically analyze. The governance of the education systems in the four contexts also does not reflect a democratic approach that involves grassroot organizations, student bodies, parents' committees and local educators and academics; education policy has mostly been a top–down approach with semi-decentralization attempts. Even when there are policies and legislations that support decentralization, local authorities and schools lack the capacities and the resources to actually implement them in a meaningful way.

Bibliography

Akkari, A. (2005). The Tunisian educational reform: From quantity to quality and the need for monitoring and assessment. *Prospects, 35*, 59–74.

Akkary, R. K. (2014). Facing the challenges of educational reform in the Arab world. *Journal of Educational Change, 15*(2), 179–202.

Allam, D. (2021). Explaining the persistence of "decentralisation" of education in Egypt. *International Journal of Educational Development, 82*, 102357.

Amira, M. (2017). Higher education and development in Egypt. *The African Symposium (TAS), 16*(1), 63–73.

Anera. (2020). *Early childhood development in Palestine, Gaza & West Bank*. Anera. https://www.anera.org/priorities/early-childhood-development/. Accessed 4 October 2020.

Assaad, R., Hendy, R., & Salehi-Isfahani, D. (2019). Inequality of opportunity in educational attainment in the Middle East and North Africa: Evidence from household surveys. *International Journal of Educational Development*, *71*, 102070.

Assaad, R., & Krafft, C. (2015). Is free basic education in Egypt a reality or a myth? *International Journal of Educational Development, 45*, 16–30.

Asali Nuseibeh, R. (2015). *Political and social exclusion in Jerusalem: The provision of education and social services*. Routledge.

Asali Nuseibeh, R. (2019). Palestinian women teachers in East Jerusalem: Layers of discrimination in the labor market. *The Middle East Journal, 73*(2), 207–223.

Asali Nuseibeh, R. (Forthcoming, 2023). Education in Palestine. In Badran, A. and Dumper, M. (Eds.), *The Routledge Handbook on Palestine* (1st ed.). Routledge.

Aubert, J., & Reiffers, J. (2004). *Knowledge economies in the Middle East and North Africa: Toward new development strategies*. World Bank.

Broco, F., & Trad, J. (2011). *Education in the Palestinian territory*. https://cemofps.org/documents/download/education_in_the_palestinian_territories.pdf. Accessed 10 September 2020.

Brownson, E. (2014). Colonialism, nationalism, and the politics of teaching history in mandate Palestine. *Journal of Palestine Studies, 43*(3), 9–25. https://doi.org/10.1525/jps.2014.43.3.9

Channa, A. (2015). *Decentralization and the quality of education*. Paper commissioned for the EFA Global Monitoring Report 2015, Education for All 2000–2015: Achievements and Challenges.

Demichelis, M. (2015). From Nahda to Nakba: The Governmental Arab College of Jerusalem and its Palestinian historical heritage in the first half of the twentieth century. *Arab Studies Quarterly, 37*(3), 264. https://doi.org/10.13169/arabstudquar.37.3.0264

El-Kogali, S. Krafft, C. (2015) *Expanding opportunities for the next generation: Early childhood development in the Middle East and North Africa. Directions in development*. World Bank.

Escardíbul, J., & Helmy, N. (2015). Decentralisation and school autonomy impact on the quality of education: The case of two MENA countries. *SSRN Electronic Journal*. https://doi.org/10.2139/ssrn.2689796

Foster, M. (1989). It's cookin' now: A performance analysis of the speech events in an urban community college. *Language in Society, 18*, 1–29.

Hanushek, E. A., Link, S., & Woessmann, L. (2013). Does school autonomy make sense everywhere? Panel estimates from PISA. *Journal of Development Economics, 104*, 212–232.

Hartmann, S. (2008). *The informal education sector in Egypt: Private tutoring between state, market, and civil society* (Working Paper 88). Department of Anthropology and African Studies, Johannes Gutenberg-Universität. http://www.ifeas.uni-mainz.de/workingpapers/AP88.pdf

Hashemi, A., & Intini, V. (2015). *Inequality of opportunity in education in the Arab region*. United Nations Economic and Social Commission for Western Asia (ESCWA).

Heckman, J., & Masterov, D. (2007). The productivity argument for investing in young children. *Review of Agricultural Economics, 29*(3), 446–493.

IAEC. (2004). *Charter of educating cities*. http://www.bcn.cat/edcities/eng/carta/charter_educating.pdf. Accessed 12 September 2021.

IAEC. (2020). *Charter of educating cities*. https://www.edcities.org/wp-content/uploads/2020/11/ENG_Carta.pdf. Accessed 12 September 2021.

Johnson, D. (2018). *Egypt's long road to education reform*. https://timep.org/commentary/analysis/egypts-long-road-to-education-reform/. Accessed 12 September 2021.

Kamel, S. (2014). Education in the Middle East: Challenges and opportunities. In N. Azoury (Ed.), *Business and education in the Middle East* (pp. 99–130). Palgrave Macmillan.

Kameshwara, K. K., Sandoval-Hernandez, A., Shields, R., & Dhanda, K. R. (2020). A false promise? Decentralization in education systems across the globe. *International Journal of Educational Research, 104*, 101669.

Karima, H. (2020). *Egypt signs $252M financing agreements for education in 2020.* https://www.egypttoday.com/Article/3/95719/Egypt-signs-252M-financing-agreements-for-education-in-2020. Accessed 12 September 2021.

Karmel, E. M., Batran, A., & Hussainy, M. (2016). Decentralizing education in Jordan: Seizing the opportunity to improve the quality and equality of Jordanian schools (Social justice in Jordan, Paper No. 2). Rosa Luxemburg Stiftung regional office.

Khuluq, L. (2005). Modernization of education in the late Ottoman empire. *Al-Jami'ah: Journal of Islamic Studies, 3*(1), 23. https://doi.org/10.14421/ajis.2005.431.23-55

Krafft, C., Branson, Z., & Flak, T. (2019). What's the value of a degree? Evidence from Egypt, Jordan and Tunisia. *Compare: A Journal of Comparative and International Education, 51*, 61–80.

Leer, J. (2016). After the Big Bang: Estimating the effects of decentralization on educational outcomes in Indonesia through a difference-in-differences analysis. *International Journal of Educational Development, 49*, 80–90.

Longworth, N. (2011). *Learning cities.* A Collection of Papers for the Xploit Girona Sessions. www.learningcommunities.eu. Accessed 12 September 2021.

Marlow-Ferguson, R. (2002). *World education encyclopedia: A survey of educational systems worldwide* (2nd ed.). Thomson Learning.

Masri, S. (2020). *Why Tunisia's once superior education system needs to reform again.* https://www.al-fanarmedia.org/2020/04/why-tunisias-once-superior-education-system-needs-to-reform-again/. Accessed 12 September 2021.

McCullum, P. (1989). Turn-allocation in lessons with North American and Puerto Rican students. *Anthropology and Education Quarterly, 20*, 133–156.

Means, A. (2014). Achieving flourishing city schools and communities: Corporate reform, neoliberal urbanism, and the right to the city. *Journal of Inquiry & Action in Education, 6*(1), 1–17.

Milovanovitch, M. (2014). Trust and institutional corruption: The case of education in Tunisia (Edmond J. Safra Working Papers, No. 44). Edmond J. Safra Center for Ethics. Harvard University.

Mullis, I. V. S., Martin M. O., & Loveless T. (2016). *20 years of TIMSS. International trends in mathematics and science achievement, curriculum, and instruction.* International Association for the Evaluation of Educational Achievement.

National Committee for Human Resources Development. (2016). *Education for prosperity: Delivering results. A national strategy for human resource development 2016–2025.* NCHRD.

Nour, O. (2013). Building child friendly cities in the MENA region. *International Review of Education, 59*(4), 489–504.

Nussbaum, M. C. (2011). *The human development approach*. The Belknap Press of Harvard University Press.

OECD. (2019). *PISA 2018 results: Combined executive summaries* (Vols. I, II, & III). OECD Publishing.

PCBS. (2017). *Selected indicators for education in Palestine by level of education and region for scholastic year 2018/2019*. http://www.pcbs.gov.ps/Portals/_Rainbow/Documents/Education2018_E.html. Accessed 4 October 2021.

Pianta, R., Barnett, W., Burchinal, M., & Thornburg, K. (2009). The effects of preschool education. *Psychological Science in the Public Interest, 10*(2), 49–88.

Pormann, P. (2006). The Arab 'cultural awakening (Nahda)', 1870–1950, and the classical traditional. *International Journal of the Classical Tradition, 13*(1), 3–20.

Robeyns, I. (2006). Three models of education. *Theory and Research in Education, 4*(1), 69–84.

Rondinelli, D., Cheeba, G., & Nellis, J. (1983). *Decentralization in developing countries: A review of recent experience* (Working Paper No. 581). World Bank.

Salehi-Isfahani, D., Hassine, N. B., & Assaad, R. (2014). Equality of opportunity in educational achievement in the Middle East and North Africa. *Journal of Economic Inequality, 12*(4), 489–515.

Sen, A. (1999). *Development as freedom*. Oxford University Press.

Sieverding, M., Krafft, C., & Elbadawy, A. (2017). *"The teacher does not explain in class": An exploration of the drivers of private tutoring in Egypt* (Working Paper No. 135). GLO.

Simão, M., Millenaar, V., & Iñigo L. (2016). *Education financing in decentralized systems enquiries into the allocative efficiency of educational investment and the effects on other dimensions of quality education policies*. UNESCO International Institute for Educational Planning.

Sobhy, H. (2012). The de-facto privatization of secondary education in Egypt: A study of 35 private tutoring in technical and general schools. *Compare: A Journal of Comparative and International Education, 42*(1), 47–67.

Stave, S. E., Tiltnes, A. A., Khalil, Z., & Husseini, J. (2017). *Improving learning environments in Jordanian public schools*. https://arddjo.org/sites/default/files/resource-files/improving_learning_environments_in_jordanian_public_schools.compressed_12_0.pdf. Accessed 12 September 2021.

Stewart, K. (2016). The family and disadvantage. In H. Dean & L. Platt (Eds.), *Social advantage and disadvantage*. Oxford University Press.

Stewart, K., & Obolenskaya, P. (2015). *The coalition's record on the under-fives: Policy, spending and outcomes 2010–2015* (Social Policy in a Cold Climate Working Paper No. 12). CASE, LSE.

Tatarsky, A., & Cohen-Bar, E. (2017). *Deliberately planned: A policy to thwart planning in the Palestinian neighborhoods of Jerusalem.* http://www.ir-amim.org.il/sites/default/files/Deliberately%20Planned.pdf. Accessed 4 October 2020.

Tatarsky, A., & Maimon, O. (2017). *Fifty years of neglect: East Jerusalem education report*. Ir Amim. http://www.ir-amim.org.il/sites/default/files/Education_Report_2017-Fifty_Years_of_Neglect.pdf. Accessed 4 October 2020.

Tibawi, A. (1956). *Arab education in mandatory Palestine*. Luzac.

Tikly, L., & Barrett, A. (2011). Social justice, capabilities and the quality of education in low income countries. *International Journal of Educational Development, 31*(1), 3–14.

UNCTAD. (2013). The Palestinian economy in East Jerusalem: Enduring annexation, isolation and disintegration (doc. no. UNCTAD/GDS/ APP/2012/1 (2013): 10).

UNESCO. (2013). *Beijing declaration on building learning cities.* Lifelong Learning for All: Promoting Inclusion, Prosperity and Sustainability in Cities. Adopted at the International Conference on Learning Cities Beijing, China. https://www.dvv-international.de/fileadmin/files/beijing_declaration_en.pdf. Accessed 22 September 2021.

UNESCO. (2020). *Key features of learning cities*. https://uil.unesco.org/lifelong-learning/learning-cities/key-features-learning-cities. Accessed 12 September 2021.

UNICEF. (2014). *Jordan country report on out-of-school children*. MENA Regional Office. UNICEF, Jordan.

UNICEF. (2018). *State of Palestine: Out-of-school children*. https://www.unicef.org/mena/reports/state-palestine-out-school-children. Accessed 4 September 2020.

UNICEF (2020). Middle East and North Africa Out-of-School Children Initiative. https://www.unicef.org/jordan/media/5501/file/OSC-Report-EN.pdf. Accessed 1 June 2021.

UNICEF. (2022). Child friendly cities initiative UNICEF. https://childfriendlycities.org/. Accessed 1 February 2021.

Unterhalter, E. (2003). *Education, capabilities and social justice*. UNESCO. https://unesdoc.unesco.org/ark:/48223/pf0000146971. Accessed 4 October 2020.

Walker, M. (2006). Towards a capability-based theory of social justice for education policy-making. *Journal of Education Policy, 21*, 163–185.

Walker, S., Wachs, T., Grantham-McGregor, S., Black, M., Nelson, C., Huffman, S., et al. (2011). Inequality in early childhood: Risk and protective factors for early child development. *The Lancet, 378*(9799), 1325–1338. https://doi.org/10.1016/s0140-6736(11)60555-2

World Economic Forum. (2018). *The global competitiveness report 2017–2018*. http://www3.weforum.org/docs/GCR2017-2018/05FullReport/TheGlobalCompetitivenessReport2017-2018.pdf. Accessed 4 October 2021.

CHAPTER 5

Higher Education and the City

Introduction

Universities and colleges contribute greatly to cities, on the economic, political and intellectual and cultural levels. Universities are deeply embedded in local economies, as they are amongst the largest landowners, developers, employers and consumers of both private and public goods in cities (Perry & Wiewel, 2015). They create a variety of jobs in management, administration, research, academia, technical support, retail, leisure, etc.—varying occupations and pay scales (Goddard & Vallance, 2013). They also create major development projects, building large scale facilities and construction projects that can be used by both the university community as well as the city community. As institutions, they also consume large amounts of goods from local and international providers.

Universities attract large numbers of students, and administrative and academic staff, who sometimes make up more than a quarter of the city's residents. These groups create housing and consumption requirements in cities, and also diversify cities, as they can come from outside those cities or even outside those countries (Goddard & Vallance, 2013). These institutions can shape the local and national culture, by opening up their facilities to the public, and by engaging the public with various cultural programs, setting the cultural tone of the city. These institutions also provide technological innovations, business innovations and research, in

R. A. Nuseibeh, *Urban Youth Unemployment, Marginalization and Politics in MENA*, Middle East Today,
https://doi.org/10.1007/978-3-031-15301-3_5

addition to evidence-based programs that have the potential to contribute to the social and economic well-being of cities and countries. They are hubs of knowledge production, whether through building human knowledge and capacities, or generating knowledge through research, publications and evidence-based development programs. However, the governance of these institutions, their access to resources and human capital and the goals and visions of their leadership, determines how open and egalitarian these institutions are, and what their impact on the local, national and global level is. These institutions have been known to produce political leaders and revolutionaries. Even in authoritarian regimes, universities have been hotbeds of revolt and dissension; students, professors and intellectuals have been a force in revolutions around countries, and regimes have been threatened and toppled by student protests (Altbach, 1989). The influence of universities on cities is great, and the relationship between universities and cities has been described as a symbiotic relationship (Klump & Bickl, 2012). It's not only cities that are transformed by and benefit from universities, it's a two-way relationship. The success of universities also depends on the quality of its urban environment, whether through university-industry partnership programs, that make it interesting for students to join these institutions, or labour market opportunities for spouses that influence staff decisions to join the institution, or enticing internships and co-placement opportunities in the city that attract students to join (Gertler, 2012). Other factors include the cultural life of the city, and transportation infrastructures. These are all pull factors that students and staff take into consideration before joining universities; as much as universities build cities, cities and their economic, political and cultural environments can also build universities.

The global neoliberal shifts have heavily influenced the relationship between the university and the city; this can be witnessed in the push for universities to assume roles as economic drivers, as well as in major financial cuts to institutions of higher education, and the flourishing of profit-making private universities, abandoning the university's civic goals. Addie (2016) in his discussion on universities in urban societies, builds on Lefebvre's (2003) conceptions of 'urban society', asking us to re-imagine the urban role of universities. The concepts he uses are mediation, centrality and difference. Addie (2016) argues that the university, which is a site of knowledge production and influence on the urban space and beyond, needs to revise the type of knowledge that is valued within the university, and how it is rendered legible for urban inhabitants. He argues that universities should "see like a city" (Magnussun, 2011: 4), by engaging in diversity of knowledge, spatial perspectives and experiences in the urban space, so students and faculty need to be trained to study the boundaries of the urban space and what is open and closed to which groups and why (Addie, 2016). In practice, he suggests for example free legal and planning consultancy to marginalized groups, and training for community organizers and support for grassroot organizations, and faculty interventions in the production of space, through urban planning (Schafran, 2015 cited in Addie, 2016). The mediation process, as explained by Addie, involves the mediatory role between abstraction and actual implementation of strategies and programs that create a more just and equitable urban space. When thinking of the centrality aspect of the "right to the city", and looking at the role of universities in cities, Addie (2016) says that universities contribute to centrality in two ways: first, they are centers in themselves, by the number of people and capital they attract, and second, they can create centrality in marginalized spaces in the city, by reaching out to excluded minorities, opening centers, creating workshops, creating attractions and giving life to these spaces. As for the concept of "difference," the university as a monumental institution in the city has the potential to act as an actor of colonization and reproduction of inequality and further elite interests, on the other hand, it has the potential to break injustice in society and promote for a more inclusive, democratic and just society. The university should review its policies and how open it is for people from different backgrounds and identities, what narratives it's promoting, who is producing which knowledge, and who is excluded from its boundaries.

This chapter examines to what extent the higher education system is democratic and egalitarian in our four cities, how it shapes the city and how the city is shaped by it. It also explores the combined "right to the city" and "capability approach" theories, when looking at higher education in our four contexts. The "capability approach" directs us to ask several questions, among them, how equal and fair access to higher education is, and what does the higher education experience mean to students in terms of quality, and the building of agency freedom. This directs our attention towards the resources, governmental policies and structures that enable or disable people's opportunities in higher education; how prepared and ready people are to access higher education, and how pre-school and school education affect opportunities in higher education.

What social norms prevail and how they shape and influence people's self-esteem or life prospects and choices in higher education? What policies are in place to remedy intergenerational injustices in society? How is the urban structure designed, what's the transportation system like and how easy can people access institutions of higher education? Is there segregation based on ethnicity, gender, social class or religion that affects people's access to resources and opportunities? What sort of higher education institutions are available and which groups are clustered in which institutions and in which disciplines?

Fair access to higher education means that people have equal opportunities and enhanced capabilities, to be able to pursue the life that they have reason to value, which is at the heart of this approach according to Sen. Looking at higher education through the lens of the "capability approach," one also needs to think of the quality of the higher education available. There needs to be a review of the quality of the curriculum and the human resources available. How relevant and empowering the learning experience is? What do governmental spending and policies on higher education suggest? Are they directed solely for market purposes and economic development, or is the aim an education targeting human development, social justice, agency, well-being and freedom, and democratic citizenship (Elkhayat, 2018; Walker, 2006)? Does it enhance and expand the different fields, whether sciences or humanities? Are the students' capabilities being enhanced by high-quality curriculums and teaching methods, that raise students' curiosities, understanding of opposing views, and freedom to critically analyze, or are they being dictated information that is outdated and does not build their agency and freedom of thinking? Also, what streams receive more governmental spending, promotion and societal prestige? Nussbaum (2010) suggests that there needs to be a focus on liberal education, as it has a role in enhancing peoples' knowledge and imagination. She commends the American system of higher education, which requires two undergraduate years of liberal education before specialization, which enhances students' knowledge and imagination and teaches them critical thinking, which is necessary for independent action and resistance to systems of oppression and bigotry. However, it also depends on the quality of the education offered. In Egypt and Jordan, for example, since the government was the major employer and it mainly asked for graduate credentials, specialization didn't matter much, so there was overrepresentation in non-scientific fields, but the nature of the curriculum, with its emphasis on rote memorization, repeating without questioning or debating did not really build capabilities, rather, it taught obedience and complicity. Another aspect that is crucial to see in higher education is how diverse, tolerant and multicultural it is (Elkhayat, 2018). Is this diversity reflected in the composition of the staff and the curriculums offered? Is higher education offering the kind of education that enhances multiculturalism and fosters a culture of open dialog? All these issues are fed from the societies that these institutions are established in, but also higher education institutions are significant societal actors that can perpetuate or break patterns of inequality and oppression in societies. Another issue this chapter will look into is the governance of institutions of higher education and the privatization drive that is increasingly being witnessed in Arab countries and how that affects the quality and equality of higher education.

University and City

Al Azhar university of Cairo is one of the oldest universities in the world. It was first founded in 970 as a mosque by Jawhar Al Siqilli—a Fatimids ruler, who conquered Egypt for his master Al Muizz li Din Allah (Rabbat, 1996). The mosque was established as a praying space in the newly founded city of Al Mansouria, which was later named Cairo *Al-Qahira,* by Al Muizz on his first visit to Egypt (ibid.). The mosque was also named the Mosque of Cairo *Jami' al-Qahira,* after the city. This religious and educational institution carried a heavy religious, political and cultural weight and was one of the major foundations of the city. The name of the mosque was later changed to Al Azhar, and the most popular explanation among Islamic scholars behind this naming is that it is the masculine name of Al Zahra, the daughter of the prophet Mohamad who was called Fatima Al Zahra—from whom the Fatimid dynasty derives its name (Rabbat, 1996). Others say that this is not the reason behind the naming, because since its founding, and for a long period of time the name of the mosque was the mosque of Cairo. Some attribute the name to the fact that the mosque was built among a number of palaces known as *al-Qusur al-Zahira* (ibid.). As Cairo became the capital of the Islamic Fatimid Dynasty, Al Azhar, as an educational and religious institution, carried a heavy political and cultural weight in the city. This symbiotic relationship between university and city was witnessed throughout different political eras in Egypt. Rabbat (1996) argues that apart from its fall from grace during the Ayyubid era, Al Azhar followed the fate of the city it was founded in, from being the capital of a religious Fatimid dynasty, to the center of an expansionist Mamluki empire, to a provincial capital of an Ottoman Empire, to the modern metropolis that Cairo became today. He says that there is a correlation between the political rulers of Egypt and their policies and ideologies and the weight and care, and even maintenance that was bestowed on Al Azhar institution, both as an architectural structure, and a religious and educational institution. Al Azhar taught Islamic studies and Arabic language, it didn't have a formal curriculum or a formal recruitment process. In fact, it was not officially recognized as a university by the Egyptian government until 1961, under Jamal Abdel Nasser, when the institution also added several secular subjects. Al Azhar has given Cairo major prominence as it attracted students and scholars from all over the world to the city. Today, the premises of the university has doubled, and it caters to over 44,000 students, and several different educational institutions are affiliated with it (Ministry of Higher Education and Scientific Research, 2022). Although for almost a century it functioned as the only university in Egypt, now it stands next to 50 other universities around the country. The University of Cairo was established in 1908 as a private university, and was named Fuad I university, when it was first founded (Reid, 2002). Then after the end of the British Occupation, which was resistant to the establishment of institutions of higher education in Egypt (to curb nationalism and resistance), the university became a state funded institution in 1925 (ibid.). The establishment of the university was the dream of prominent Egyptian academics, who hoped to create a modern secular institution of higher education that would complement Al Azhar religious studies, and offer an alternative kind of higher education (ibid.). Although the university had shaky beginnings with limited resources, it is now considered as a major university in the region enrolling around 207,000 students and hiring around 14,000 staff (Ministry of Higher Education and Scientific Research, 2022). The university has a heavy economic, political and cultural weight in the city.

Similar relationships between university and city can be seen around the Arab world. Tunis for example is the home of one of the oldest universities as well. Az-zaytouna University, which was also previously a mosque established by Ubayd Allah ibn al-Habhab in 731, is one of the oldest Islamic learning institutions (Deeb, 1995). The fate of the educational institution was also intertwined with the fate of its city. In the realm of the Hafsid dynasty in the thirteenth century, Tunis flourished, becoming the capital of Ifriqiya[1]; Az-zaytouna, as a major religious and educational institution in the city also flourished, becoming one of the major centers of Islamic learning. It attracted students and scholars from around the Islamic world, so the university contributed to the importance of the city of Tunis in the Islamic world. One of its prominent alumni is Ibn Khaldoun, who was a major sociologist, historian and philosopher (Charnay, 1979). In 1875, over a millennium after the establishment of Az-zaytouna University, Al Sadiqieh College was established as the first modern era college, with a multicultural multilingual curriculum, which was even reflected in its architecture that contained Arabic, Turkish and European elements (Essabiti, 2016). The majority of college students continued their studies abroad at western and mainly French universities. It is said that the college has educated Tunisia's prominent scientists, engineers and political figures. The majority of the Destour party is said to have studied at Al Sadieqieh, among them the prominent ex-president Habib Bourguiba (ibid.). During the French colonization in 1945, the Institute of Higher Studies was established which was affiliated with the university of Paris. This paved the way for the establishment of De Ecole Normale Superieure, an institution of higher education in Tunis in 1956, hence connecting Tunisian higher education systems with France. So prior to Tunisia's independence from French colonization the higher education had three pillars: Islamic religious studies offered at Az-zaytouna; French Western education offered though Ecole Normale Superieure; and civic bilingual education offered through Al Sadiqieh college (Fryer & Jules, 2013). When Tunisia gained its independence and Bourguiba took office, he had a policy of "Tunisification" and "modernization" of institutions of higher education (ibid.). In his strategy to diminish the role of Islamic religious institutions in the country, at the same time, maintaining and respecting the cultural and religious significance of Az-zaytouna, regionally and in the Islamic world, Bourguina brought Az-zaytouna and Ecole Normale Superieure together and established Universite' de Tunis, which is known today as University of Tunis El Manar (ibid.). The university has around 33,000 students today and stands next to 13 other public universities in Tunisia (Université de Tunis El Manar, 2022; Gdoura, 2019).

[1] The area of what constitutes today parts of Libya, Algeria and Tunisia.

Amman's more recent history has a similar connection between city and university. The University of Jordan was founded in 1962 in an area that was considered just outside Amman at the time. It was founded by a royal decree at a time when the West Bank was still under Jordanian rule, and Jerusalem was the culturally richer city (Cantini, 2012). In a move that was politically motivated to give a cultural and economic weight to Amman, the university was established in the East Bank (ibid.) The university started with 167 students and 8 academic staff and today has around 49,000 students and around 1,500 academic staff (University of Jordan, 2022). The University District is one of the most populated and developed areas of Amman today; so the university did manage to bring a cultural and economic weight to the city. It has also managed to graduate the majority of the country's entrepreneurs and politicians. It stood as the only university in the country for 27 years until the Ahliyya Amman private university was established in 1989; that's why the University of Jordan was labeled as the mother university. Today, there are several private and public universities mainly in the city of Amman. The majority of the staff during the foundation years of the university were Palestinians, given that the majority of the population back then was Palestinian, and they were overrepresented in urban spaces and had higher levels of education (Cantini, 2012). After the events of black September 1970, Palestinian presence in the public and state spheres started diminishing, and they were more present in private institutions. This worked in their favor with the wave of global privatization, as they were less dependent on government jobs and benefits, however, this fueled further political tension with Jordanian East Bankers (Susser, 2021).

Palestinians have tried to establish their own institution of higher education in Jerusalem dating back to 1931 when the first and only Islamic summit was held in Jerusalem. One of the results of the summit included the establishment of an institution of higher education in Jerusalem. Palestinian academics' requests to build their own university were denied, while Israelis were granted permission by British mandate authorities to build their Hebrew University as early as 1925 (Demichelis, 2015 cited in Asali Nuseibeh, forthcoming). In 1957, when East Jerusalem was under the governance of Jordan, a board of trustees was formulated in Kuwait to work on the establishment of a Palestinian university in Jerusalem, the board hired an executive committee headed by the mayor of Jerusalem back then, Rawhi Al khatib (Al-Quds University, 2021). The first stone to build the university was placed in 1966 in a ceremonial event by King Hussein of Jordan. However, the events of the 1967 war and the occupation and annexation of East Jerusalem, put a halt on Palestinian dreams of having their own institution of higher education in Jerusalem. Then in 1977, the college of Islamic studies opened in Beit Hanina—a village north of Jerusalem. Two years later, the first building of the university was established in Abu Dis, and the college of science was opened just to be closed again by Israeli authorities. The relationship between Israel and the Palestinian Al Quds University has a long and symbolic history of Israel's approach towards Palestinian identity, culture and presence in the city of Jerusalem. Since its establishment, Israel has sought to repress the progress and expansion of Al Quds University; reflected by its attacks on students, faculty and university premises (Hoyle, 2015), limiting the number of foreign academics visiting Palestinian territory (Griffith et al., 2022) and not accrediting most of the Al Quds University degrees (Abramson, 2016). However, with all the limitations and restrictions placed on it, the university has managed to create 120 undergraduate and postgraduate programs, taught through fifteen colleges, and has a student enrollment today of 12,000 students (Al-Quds University, 2021).

Access and Equality

> The desire for a college education is not a brute fact of nature, but is shaped by what you think about yourself, what amount of self-esteem you are led to have by your society, what your society tells you about the opportunities that are likely to be open to you, and so on. (Nussbaum interviewed in Pyle, 1999: 244 cited in Walker, 2006: 101)

In Jordan, Tunisia and Egypt, access to public universities is centralized. The determining criterion for admission to public universities is the grade achieved in the national examination at the end of the secondary school, which is *Tawjihi* (The General Secondary School Certificate Examination) or its equivalent, in the case of Egypt and Jordan, and the baccalaureate, in the case of Tunisia. The students submit their applications to the higher education coordination office, including their secondary school examinations transcripts, and their preferences for universities and subjects. Then, the coordination office distributes the students among public universities, based on their scores, and partially, based on their desire of what to study and where. At some universities, there's the option for students to pay extra fees and study their desired field, and are treated as international students. Students can also join private universities and colleges to study the field of their choice, but they need to be able to afford the fees of private institutions. Higher education in the Arab world is state-controlled and highly centralized, with the two exceptions being Palestine and Lebanon, where political conflict contributed to the fragmentation of state power, hence undermining the ability of the state to develop a state-controlled higher education system (Buckner, 2011).

In Egypt, Tunisia and Jordan, ministries maintain tight control over curricula, admissions and recruitment. Ministries use computerized systems to channel students into disciplines and institutions. This means that many students find themselves studying disciplines they do not desire, and this partly explains why many students never complete their degrees. This centralized system gives only a minority of students the choice to enroll in the disciplines in which they are interested in, and those students are mainly supported by their parent's resources, in order to secure the most "prestigious" positions in institutions of higher education. There is also the parental and societal pressure in the Arab world, where being a medical doctor or an engineer has more societal value than being an artist or a social scientist.

> We have something called the top colleges; medicine and engineering. Once I was in public transportation, and a guy saw me sitting with a bag, and he said to me are you studying or working, and I said I am a student, and then he said to me engineering or medicine (laughter). As if there is no other option for studying; you either study medicine or engineering.[2]

The majority of students who score on the medium to lower spectrums on their exams, when graduating from school, will have to consider taking a place in the "remaining" disciplines that have been assigned to them (Milovanovitch, 2014). Those faculties are usually less costly, and acquiring a position in those faculties is considered less "prestigious" in society, examples include humanities, education and social sciences, which are mostly dominated by female students. Being forced to study a field not of your choice, limits the motivation of students and lowers their morale and academic ambitions (Milovanovitch, 2014).

[2] Interviewee Salim (Pseudonym), 19, Muslim, Male, Political Science Student, Cairo.

Another "dumping grounds" for students that no one wants, are vocational schools (Wilkens, 2011: 8). Post-secondary vocational schools are dysfunctional in Arab countries; with poor management, resourcing, curriculums and methods of teaching.

In Egypt and Jordan, private higher education started growing in the nineties. In Tunisia, private education started expanding in 2001 (Gdoura, 2019). As higher education in those countries evolved from being available only to the privileged elite, into mass access, governments were not able to fulfil the growing demand, and so private institutions started to grow to accommodate for the growing demand in higher education. High birthrates in combination with expansions in primary and secondary school education have led to high number of school graduates, which higher education institutions cannot absorb. Ministries of education in Arab countries find themselves in a tricky position as they publicly advertise for higher success rates in secondary education exams, while at the same time cannot cope with the large numbers of people wanting to join higher education (Coffman, 1996).

The massification of higher education in Egypt started in the 1960's under Jamal Abdel Nasser, with the promulgation of higher education as a free right in 1962; a guaranteed employment scheme of graduates was also initiated in the same year, which led to further demand for higher education (Barsoum, 2014). The public sector had been the dominant employer of graduates, with 70% hired in the public sector in Egypt in 1965 (Assaad, 2014). This policy of providing free education from primary through higher education remains to this day, but the promise of a guaranteed government job started diminishing since the 1980's (Abdelkhalek & Langsten, 2020). The government realized that it is not able to absorb the high demand for higher education and government Jobs. To absorb the high demand for higher education without increasing cost on the government, policymakers followed two policies. The first policy was allowing private for-profit higher institutions to operate in Egypt, and the second was diverting students towards vocational training (Abdelkhalek & Langsten, 2020). The two policies followed did not necessarily improve the quality of higher education in Egypt, nor contribute to equality of access (ibid.).

Even with the expansion of private universities, the majority of Egyptian students still study in public institutions of higher education. In 2017, there were 25 public universities in Egypt, serving 2.2 million students and 22 private universities serving 138,000, meaning that the higher education public sector is still dominant (CAPMAS, 2017).

The second policy followed in Egypt to reduce the pressure on institutions of higher education is diverting students from academic education to vocational secondary education, where the majority of students end their formal education upon graduation of high school (Abdelkhalek & Langsten, 2020). Secondary tracking determines a student's likelihood to continue to higher education. Students with higher grades get tracked to the academic stream and those with lower grades get tracked to the vocational stream. Secondary tracking is based mainly on the score of the middle school exam, and in some cases the aspiration of the student and their parents, which means the ability of the parents to pay for private tutoring, etc. (Cupito & Langsten, 2011. It is also highly correlated with the family's wealth; 75% of children of families from the wealthiest quintile join the academic track in Egypt, in comparison to 20% of the poorest quintile (Cupito & Langsten, 2011 cited in Abdelkhalek & Langsten, 2020). While the children from poor socio-economic status are tracked to the less prestigious vocational training track, wealthy children continue in the academic track and proceed to higher education. The financing of higher education is biased against the poor, since most of the public spending on higher education goes to the richest students who have higher accessibility to these institutions, only 9% of youth from the poorest households access higher education in Egypt, in comparison to 80% of youth from the richest households (Assaad, 2013; Fahim & Sami, 2011). Institutions of higher education award the cultural capital of wealthier students, which is manifested in language, skills and knowledge, that their parents provide them with (Bourdieu & Passeron, 1977). This can range from school choice, to after school activities, to private tutoring which is prevalent in Egypt, Tunisia and Jordan. The absence of a strong public primary and secondary education system means that parents private spending on their children's education will tip the scales in their favor when it comes to access to higher education.

Parents' education, especially the mother's education, is also an important determinant of access to higher education (Krafft & Alawode, 2018). In the three countries Tunisia, Egypt and Jordan, the fathers' profession also affects the children's chances of attaining higher education. Students with fathers working in the public sector have a higher likelihood of attaining higher education, especially in Egypt, which has the greatest disparity between higher education attainment for those with fathers working in the private and public sector (ibid.). Since social inequality prevails in access to police and army positions, this perpetuates a circle of inequality in Egypt. Although not proven officially, but having influential relatives and acquaintances affects people's chances of joining the police academy, or any of the five military colleges in Egypt (Amira, 2017). There's also a residency influence on people's chances of attending higher education, with people in capital cities (Cairo, Tunis, Amman) more likely to access higher education (Krafft & Alawode, 2018). Discriminatory policies banning students from access to specific faculties adds to inequality of opportunity. For example in Cairo, the decision of the Supreme Council of Universities to deny students from provincial areas to access specific faculties at Cairo University, like Media and Political Science, claiming that they have equivalent faculties in their governorates, while exempting the children of government officials from this rule, further fuels inequality, because the students' denied access miss the opportunity of being exposed to better quality teaching, and facilities that can give them a better educational experience (Amira, 2017). Therefore, to remedy inequities in education, the structural frameworks that generate these inequities need to be addressed. Undemocratic societies with unequal economic and social circumstances can create atmospheres where people have unequal opportunities and capacities to choose the life that they have reason to value.

The geographic location of Egyptians also determines their opportunities in accessing higher education, despite their abilities or ambitions. The typical impact analysis of geographic location on higher education is usually concerned with rural/urban disparities. However, researchers are now seeing that inequality is very high within urban areas as well (Tadamun, 2018). The unequal distribution of resources, services and transportation routes can impact people's access to higher education and employment prospects. Overcrowding in poor neighborhoods, and flawed statistics that do not take into account informal housing in Cairo, means that there is a limited number of secondary schools in poor neighborhoods. This partly explains why students residing in poor neighborhoods are less presented in secondary education, in comparison with wealthy students, and why they have an even lower probability of attending higher education. Tadamun (2018) provides an example, where in one of greater Cairo's informal neighborhoods, Izbit Khayrallah—which is home to an estimated 650,000 people, there is only one primary school and no secondary schools. In comparison, the district of 1st New

Cairo, which is one of the wealthiest in Cairo, with a relatively small population of 27,000, has 49 secondary schools (ibid.). Manshiet Al Nasser, another neighborhood with high levels of poverty, and an estimated population of 260,000, has only one secondary school. This factor, added to other variables such as parents' income and education, can explain why 48% of 1st New Cairo residents have a university education, compared to only 3% of Manshiyat Al Nasser residents (ibid.).

The massification of education in Tunisia started in the 1990's, during Ben Ali's Era. During Bourguiba's Era (1957–1987), the education system mandated a national exam at the end of primary school, which determined eligibility for admission to secondary school. This exam effectively blocked 50% of students continuing to secondary school; those who did not pass joined vocational schools (Masri, 2020). In 1991, Ben Ali annulled the secondary school entrance exam and made school compulsory until the age of 16, he also made it easier for secondary school graduates to enter university, so the number of graduates per year increased significantly from 5,552 in 1990 to 52,130 in 2019 (Assaad et al., 2017; MoHE, 2020). Today there are 204 institutions of higher education in Tunisia, of those, there are 13 public universities (MoHE, 2020). There are 74 private institutions of higher education accommodating 39,535 students (ibid.). The majority of students, numbering 234,029, are studying in the public sector (ibid.). Again, although the higher education system in Tunisia is mainly public, children of wealthy families still have a higher chance of attending higher education, and are clustered in more "prestigious" institutions and specializations, as they are better prepared through private tutoring, and are able to access higher quality schools in more affluent areas.

Jordan has been leading the privatization model of higher education, compared to Egypt and Tunisia, with the highest levels of enrollment in private institutions—27% of students (Assaad et al., 2016). In Egypt and Tunisia, enrollment in private institutions remains limited, but is on the rise. Privatization of higher education went hand in hand with the economic privatization policies in the Arab world. This wave of privatization was influenced by a global policy of privatization, which was pushed by donor organizations (Al-Atari, 2016). Advocates for the privatization of higher education argue that the private sector is more efficient, better managed, more fluid and more responsive to changes in the modern world; therefore, it is able to generate resources more efficiently. They describe the public sector as bureaucratic, unresponsive and wasteful (ibid.). This positive view of privatization stems from a globalization ideology that places education as a key element in building human capital for a competitive global market (ibid.).

In Egypt, Tunisia and Jordan, public spending on higher education did not go hand in hand with the increase in demand. Barsoum (2017) explains that the massification of higher education presents a major challenge to state budgets, and because it's not being met, its resulting in overcrowded lecture halls, outdated facilities and supplies, and faculty with less time for research or effective teaching, which paves the way for a bigger role to be played by private institutions. Private universities in Egypt, Jordan and Tunisia are regulated by the state, but enjoy greater autonomy than public institutions. In the three countries, private higher education is a fallback when access to public education is not possible, and it is normally an option for families who can afford to pay the fees.

> There are these for-profit private universities that only, sorry to say this, the losers get in those. For example, for a specialization that requires 95 in a public university, you can go into the same specialization on 75.... And here there is also classism. You have two kinds of people, the kind that says my dad is going to pay, because he's so rich, and the student is not that smart to go into top ranking universities. There's also the other kind where a father thinks I want to see my son become an engineer, and so he works day and night to get the money for that.[3]

Some argue that despite the boom of private universities, they have not managed to meet the high demand for higher education or improve the quality. With some exceptions, the newly established private institutions are viewed as further undermining quality and equity of higher education in Arab countries, and producing graduates at high personal cost, without the knowledge and skills (Wilkens, 2011). Assad et al. (2016) have found that the claim that "privatization is the solution for the problems in higher education sector" is flawed, since it does not guarantee better educational processes or quality. They argue that with more autonomous governance processes, incentives and resourcing, it is possible for public institutions to perform better, or at least as well as private institutions. Barsoum (2017) also explains that in the context of Egypt, privatization did not

[3] Interviewee Salim (Pseudonym), 19, Muslim, Male, Political Science Student, Cairo.

improve the quality of higher education. She explains that these institutions operate in a social field that valorizes credentials beyond learning, which discourages their potential for innovation, especially lower-tier private institutions, that are the choice of lower-income students who did not have the resources to be prepped for prestigious institutions, whether public or upper tier private ones, so they look for a "quick and easy" way to acquire university credentials in a "credentials fetish" society (Barsoum, 2017: 105).

> I did a masters degree at a private university, and I don't think I learned much. I just needed the degree to find a job.[4]

Private institutions mirror some of the ailments of public institutions in relation to rote memorization, as the key pedagogical approach, with little student-focused learning (Barsoum, 2014). Private institutions find themselves in a position, where they have to provide affordable education, make a profit and deal with a student body that received lower scores in the secondary education completion exams, which puts these institutions at a disadvantage in terms of performance and quality of educational output (Barsoum, 2017). This partly explains why graduates from private higher institutions do not fare differently in the labor market compared to their peers from public higher institutions (Barsoum, 2014). That said, some private institutions in Egypt tend to have more innovative pedagogy and are more accountable than public institutions, but this is not always the case when comparing public and private institutions; in the case of Jordan, the opposite is true, pedagogical and accountability practices are superior in public institutions (Assaad et al., 2017).

Researchers, supporting the privatization of tertiary education, have argued that decentralization is a strong predictor of enrollment expansion (Schofer & Meyer, 2005 cited in Buckner, 2011). This claim is supported by trends in the four countries we are examining. Enrollment in higher education in Egypt is 35.16%, in Tunisia is 31.85%, in Jordan is 34.32%, and in Palestine enrollment in tertiary education is the highest at 43.24% (UNESCO, 2017). In Palestine, there are 16 universities and 34 colleges (MoHE, 2021). Of those institutions, 11 are funded by the government, and the remaining are funded by private non-profit and

[4] Interviewee Ahmad (Pseudonym), 30, Muslim, Male, Digital Marketing Employee, Tunis.

for-profit bodies. Palestinian students' enrollment at Israeli universities is also increasing, with 17% of the total student body being Palestinian, this number includes Palestinians with an Israeli citizenship and residency permits (Council for Higher Education, 2021). Part of why higher education is more in demand in Palestine, is the destruction of the agriculture sector. The confiscation of land and water by the Israeli Occupation, as well as Israeli settler violence against Palestinian farmers, and movement restrictions, have made jobs in agriculture and farming difficult, pushing young Palestinians to pursue jobs as professionals in cities, which demands higher education degrees.

The history of higher education in Palestine is relatively new. As previously mentioned, Palestinians were denied the licensing to establish institutions of higher education during the British Mandate (Demichelis, 2015). Universities in Palestine were established in the 70's under occupation. When the Palestinian Authority took the responsibility for the education sector in 1993, the higher education system expanded. However, even after the establishment of Palestinian Authority institutions of higher education, they were functioning within a context of occupation, resulting in restrictions around political freedom of speech, movement restrictions and budget limitations, due to the poor economic environment in Palestinian territories. The siege imposed on Gaza by Israel also limits Gazan students' abilities to study in the West Bank, East Jerusalem and Israeli territories, and vice versa.

In Jerusalem, Palestinian students are leaning more towards studying at Israeli universities and colleges (Halabi, 2022). The reason being their inability to find employment in Jerusalem, or the Israeli labour market, with their Palestinian university and college degrees. They end up joining second tier Israeli institutions, due to their weakness in the Hebrew language and their inability to get high scores in Israeli entrance exams and requirements. Also, crossing checkpoints to study or work at Palestinian institutions adds to their transportation costs and time, as well as the added humiliation and suffering passing checkpoints.

> This was a huge suffering for me and maybe that's what makes me reluctant to find a job in Ramallah. I used to spend hours on checkpoints and especially in the winter with everything you are wearing it rings in the

> security machine, and you have to take off your boots. Honestly, it's so humiliating.[5]

This has also resulted in the mushrooming of several for-profit colleges and schools that promise to provide the students with the needed degrees to be able to penetrate the Israeli labour market and Israeli institutions. They charge high amounts of fees and flourish on people's lack of knowledge on how to access Israeli institutions, in addition to their weak command of the Hebrew language.

> Many times, people who study at those private colleges pay so much and then they don't find jobs in their fields and end up working other things. The girls who studied with me are all working in other things…My salary is so low that I still haven't managed to gather the money I paid for my higher education, not even half of it.[6]

The Quality of Higher Education

Once upon a time, Arab countries such as Egypt, Tunisia and Morocco were leading the world in the establishment of institutions of higher education. And although over the centuries, higher education has been expanded and opened for the masses in Arab countries, the quality of higher education today is among the lowest in the world, with higher education in Egypt ranking 100 out of 137 in the 2018 global competitiveness index, Tunisia ranked 82 out of 137 and Jordan, in a slightly better position at 63 out of 137 (Schwab, 2018), Palestine is not included in the index. Rote memorization, without critical thinking and debating, is prevalent in the majority of Arab schools and universities. The Global competitiveness report for 2018 shows that Egypt ranked 123 out of 141 in using critical thinking teaching methods, followed by Tunisia which ranked 99 out of 141, followed by Jordan, which ranked better at 37 out of 141 (ibid.).

> I didn't see myself studying in the public higher education system. I didn't see myself in the Egyptian system. Although I have managed to excel in this system in the secondary level, but I would just memorize things and

[5] Interviewee Salma (Pseudonym), 21, Muslim, Female, Marketing Student, Jerusalem.

[6] Interviewee Farah (Pseudonym), 24, Muslim, Female, Shop Employee, Jerusalem.

> then go and recite them in the exam and this is a big problem. The system is just turning me into a machine, without thinking without skills without nothing unfortunately.[7]

In the MENA region, postcolonial state-led developments included expansions of the higher education system; graduates of the higher education system would be mainly hired by the civil service, so higher education credentials were more important than people's actual skills (Assaad et al., 2017). Today, with the cuts in government jobs and the discrepancy between the jobs available and entrants to the labour market, a university degree is no longer a guarantee to enter the labour market.

Assessments made of higher education in the Arab world, have been strongly associated with the economy and preparing the youth for the labour market, and providing youth with the skills "needed." This discourse is market-oriented, and concerned with how employers can make use of available human resources. It is a purely economic perspective, which does not take into account the learners experience and what the educators see as crucial for enhancing human agency and capabilities (Walker, 2003). A capabilities' pedagogy means deep participation and engagement of the learners. It also means the recognition of diversity and cultural difference, hence recognizing the value of the cultural resources students bring to learning (ibid.).

There are several factors distorting the functioning of both public and private higher education institutions in the Arab world. First, as we mentioned before, the demand for higher education exceeds the number of places available. Second, the government is still a major employer of higher education graduates in Egypt, Jordan and Tunisia, shifting the priority from quality learning, to focusing on getting the credentials needed to obtain public sector jobs. This issue of focusing on credentials has dis-incentivized students from seeking a higher quality educational experience. Another issue affecting the quality of the educational experience at Arab universities, is the academic preparedness of secondary school students, who arrive at university lacking essential language, science, math and critical thinking skills that are needed to study at the tertiary level, which means that universities need to spend more time and resources to address these deficiencies (Wilkens, 2011). Another major issue, affecting the quality of higher education, is the lack of clear focus

[7] Interviewee Salim (Pseudonym), 19, Muslim, Male, Political Science Student, Cairo.

in research priorities and strategies, and limited research resources at the institutional level; investment on research and development as a percentage of GDP in our four contexts is very low. In Palestine, it is 0.49%, followed by Tunisia at 0.6%, followed by Jordan at 0.71, and then Egypt at 0.72, while the OECD average is 2.47% and Israel has the highest percentage at 4.9% of its GDP spent on research and development (OECD, 2019; UNESCO, 2018a). In addition, the teaching cadres, who are mostly graduates of the same weak system, lack the awareness of the importance of good scientific research. The number of researchers per million in Tunisia stands at 1,772, in Egypt it stands at 687, in Jordan 596 and in Palestine 575, whereas for OECD members the average is 4,063, and Israel has the highest figure at 8,255 (UNESCO, 2018b).

A weak research environment, which is the result of a lack of vision, strategies, resources, human capital and political power, leads to low research outputs such as publications, inventions and patents. According to SCImago, the number of scientific publications in Palestine were a mere 1,345 for the year 2020, for Jordan it was 6,849, and for Tunisia it was 8,890. As for Egypt, in the last decade, it has worked intensively on increasing its publication outputs, and has succeed, becoming the top producer of peer-reviewed journal publications in Africa, and in the 30th position globally (Sawahel, 2021). However, there is still space to improve the quality of the publications they produce and the contribution of research outputs. Egypt has one of the highest number of retractions of articles in the region. Its retraction rate is also more than twice that of the US, and three times that of the UK (Plackett, 2019). Retractions happen as a result of faked data, plagiarism, or any circumstances that have caused the research results to be questioned (ibid.). High-quality research outputs require a thriving research environment with highly qualified cadres, vision, strategy and resources.

Another major issue with regards to the quality of education available, is the weak digital infrastructure at the majority of Arab universities (apart from the Arab Gulf). This was felt strongly during the COVID-19 pandemic and the shift to online learning. Added to the problem of digital infrastructure, availability of devices and connection costs, the language of instruction is mostly Arabic, and Arabic has made limited inroads on the digital landscape (Guessom, 2006). Another major problem related to online learning and research, is that it is very individualistic and independent; the skills of life-long learning and independent research are not traits that the education system in the Arab world prepares the students for (ibid.). The students are not encouraged to be active independent learners, instead they are mostly taught to be passive receivers of knowledge (ibid.). The forced shift to online learning during the pandemic, has also amplified inequality amongst Arab students, with some students having the ability to be in a private space with high-speed access to the internet, with the availability of devices at home, and others deprived of those privileges. The sharp "digital divide" between students pushed some professors in Tunisia to call for an end to online learning (Faek & Abed Al Galil, 2020). The pandemic has only revealed to academics and policymakers in the Arab world, how weak the technological literacy and digital infrastructure is in their countries. This is a major problem, as the world has already entered the technological revolution, and apart from sporadic initiatives, technological advances have been lagging in the Arab World. Arab entrepreneurs have launched programs to support students online such as "Nafham" in Cairo, which followed the model of Khan Academy in providing students with videos and texts to help them circumvent private tutors (Sadek, 2012; Nfham, 2022), and "Edraak" in Amman, which is an online platform offering courses in Arabic in various science and arts subjects (Edraak, 2022). However, these initiatives remain limited in number.

Governance of the Higher Education System

Another factor affecting the quality of education at Arab universities is the lack of student involvement in the teaching/learning processes. Students' knowledge and experiences, are not taken into account during the teaching process, while students' feedback on lecturers and courses, is also hardly valued. For example, in Egypt, both private and public universities fail to allow students to provide feedback on the performance of instructors, or the learning experience, either through exit surveys or interim satisfaction surveys (Barsoum, 2014). The lack of democratic values in the teaching/learning environment is a reflection of the authoritarian governance of higher education systems. This governance approach does not only disempower student bodies, it also disempowers staff, and reduces the influence of civil-based organizations on higher education systems.

Governance of the institutions of higher education is another issue affecting the quality of the services available. In Tunisia, Egypt and Jordan, higher education institutions are run as extensions of the authority of the state, with no policies that enable autonomy and transparency of tertiary educational institutions (Wilkens, 2011). Slow governmental bureaucracies control the micro details of these institutions, from curriculum design and hiring processes, to the approval of new degrees, etc. (Wilkens, 2011). The central government is involved in the nomination of presidents of public universities, and in many cases, the university deans, which means that there's no transparency in the hiring process of higher education leaders, and they cannot be held accountable to the public or to the institution they are working for (Wilkens, 2011). For example, in an attempt to limit faculty and student participation in governance and to contain opposition groups, Egypt passed law 142 of 1994, which added deans to the list of senior university officials who are appointed by the minister of higher education (Mazawi, 2005). Universities in Jordan have a greater level of autonomy than those in Egypt (Assaad et al., 2016). For example, universities in Jordan have a board of trustees by law, which is responsible for strategic planning, setting budgets, the hiring process of vice presidents and deans and the setting of university tuition fees (ibid.). However, universities in Jordan face other types of restrictions coming from The Council of Higher Education, which appoints presidents of both public and private universities, and allocates budgets to public universities. This Council is chaired by the minister of higher education.

The state's political subordination of higher education institutions, emphasizes the authoritarian mode of control, that prevents the emergence of an authentic and independent academic leadership that serves the demands of the people democratically (ibid.). The state's political subordination of higher education, also goes hand in hand with the governing bodies' political and economic neoliberal policies, that seek to realign higher education objectives with privatization polices and labour market demands, rather than having a vision of enhancing people's capabilities and agency freedom. Censorship and limitations on freedom of speech at universities ensures that both staff and students remain compliant with government narratives and agendas.

As nations shift from agricultural and industrial based economies, to services and knowledge based economies, specialization and higher education is becoming more of a prerequisite for success in the labour market. Therefore, the rife inequalities in access to these institutions, and the quality of training and the specializations they are receiving, can have a great impact on their career trajectories and economic and social wellbeing. It is therefore important, not to limit discussions around higher education, to a "human capital" perspective, that emphasizes only measurable indicators of higher education performance, in terms of engaging labour markets, employability and economic returns of graduates (Mazawi, 2005), as it ignores key questions on how education can enhance people's agency and freedom, and how faculty and students can participate in, and influence, education governance.

Conclusion

This chapter explored higher education through the concepts of the "right to the city" and the "capability approach". It explored the two-way relationship between the university and the city; while institutions of higher education contribute highly to cities, on the economic, political and intellectual and cultural levels, the cities' infrastructure, political and economic stability also increased demand and attracted cadres to their universities.

The importance of institutions of higher education at the city and the country levels, in terms of their contribution to the economy, research and technology, shaping ideology and influencing political discourse, means that these institutions can have the power to either reproduce hegemonic powers and entrench inequalities, or be the power to change and give voice to the marginalized. Therefore, their governance is as important as the governance of cities. The political subordination of institutions of higher education to authoritarian governments, lack of transparency, and lack of independence greatly affects these institutions' ability to work for the greater good of the people.

Bibliography

Abdelkhalek, F., & Langsten, R. (2020). Track and sector in Egyptian higher education: Who studies where and why? *FIRE: Forum for International Research in Education, 6*(2), 45–70.

Abramson, E. (2016). Israeli forces launch another raid on Al-Quds University in East Jerusalem. https://www.endangeredscholarsworldwide.net/post/israeli-forces-launch-another-raid-on-al-quds-university-in-east-jerusalem. Accessed 20 May 2022.

Addie, J. D. (2016). From the urban university to universities in urban society. *Regional Studies,* 51(7), 1089–1099.

Altbach, P. (1989). Perspectives on student political activism. *Comparative Education, 25*(1), 97–110.

Al-Atari, A. (2016). Internationalization of higher education with reference to the Arab context: Proposed model. *IIUM Journal of Educational Studies, 4*(1), 6–27.

Al-Quds University. (2021). *Al-Quds at a Glance*. https://www.alquds.edu/en/home/al-quds-at-a-glance/. Accessed 6 December 2021.

Amira, M. (2017). Higher education and development in Egypt. *The African Symposium (TAS), 16*(1).

Assaad, R. (2013). Equality for all? Egypt's free public higher education policy breeds inequality of opportunity. In A. Elbadawy (Ed.), *Is there equality of opportunity under free higher education in Egypt?* (Arabic) (pp. 83–100). Population Council.

Assaad, R. (2014). Making sense of Arab labor markets: The enduring legacy of dualism. *IZA Journal of Labor & Development, 3*(1).

Assaad, R., Badawy, E., & Krafft, C. (2016). Pedagogy, accountability, and perceptions of quality by type of higher education in Egypt and Jordan. *Comparative Education Review, 60*(4), 746–775.

Assaad, R., Krafft, C., & Salehi-Isfahani, D. (2017). Does the type of higher education affect labor market outcomes? Evidence from Egypt and Jordan. *Higher Education, 75*, 945–995.

Attari, A. (2016). Internationalization of higher education with reference to the Arab context: Proposed model IIUM. *Journal of Educational Studies, 4*(1), 6–27.

Barsoum, G. (2014). *Aligning incentives to reforming higher education in Egypt: The role of private institutions*. Economic Research Forum.

Barsoum, G. (2017). The allure of 'easy': Reflections on the learning experience in private higher education institutes in Egypt. *Compare, 47*(1), 105–117.

Bourdieu, P., & Passeron, J. C. (1977). *Reproduction in education, society and culture*. Sage.

Buckner, E. (2011). The role of higher education in the Arab state and society: Historical legacies and recent reform patterns. *Comparative and International Higher Education, 3*(1), 21–26.

Cantini, D. (2012). Discourses of reforms and questions of citizenship: The university in Jordan. *Revue Des Mondes Musulmans Et De La Méditerranée, 131*, 147–162.

CAPMAS. (2017). *Statistical yearbook*. Central Agency for Public Mobilization and Statistics.

Charnay, J. P. (1979), Economy and religion in the works of Ibn Khaldun. *The Maghrib Review, 4*(1), 6–8.

Coffman, J. (1996). Current issues in higher education in the Arab world. *International Higher Education* (4).

Council for Higher Education. (2021). *Accessibility of higher education in the Arab sector*. https://che.org.il/en/הנגשת-ההשכלה-הגבוהה-לחברה-הערבית-2/. Accessed 5 May 2022.

Cupito, E., & Langsten, R. (2011). Inclusiveness in higher education in Egypt. *Higher Education, 62*(2), 183–187.

Deeb, M. (1995). Zaytūnah. In John L. Esposito (Ed.), *The Oxford encyclopedia of the modern Islamic world* (pp. 374–375). Oxford University Press.

Demichelis, M. (2015). From Nahda to Nakba: The governmental Arab college of Jerusalem and its Palestinian historical heritage in the first half of the twentieth century. *Arab Studies Quarterly, 37*(3), 264.

Edraak. (2022). *Online courses in Arabic*. https://www.edraak.org/en/. Accessed 24 May 2022.

ElKhayat, R. (2018). The capabilities approach: A future alternative to neoliberal higher education in the MENA region. *International Journal of Higher Education, 7*(3), 36.

Essabiti, A. (2016). Tunisian *"Sadiqiah"... Icon of science, struggle and statesmen*. https://www.aa.com.tr/ar/الثقافة-والفن/الصادقية-التونسية-أيقونة-العلم%D9%90-والنضال-ورجالات-الدولة/552888. Accessed 6 December 2021.

Faek, R., & Abed Al Galil, T. (2020). *The shift to online education in the Arab world is intensifying inequality—al-fanar media*. https://al-fanarmedia.org/2020/04/the-shift-to-online-education-in-the-arab-world-is-intensifying-inequality/. Accessed 24 May 2022.

Fahim, Y., & Sami, N. (2011). Adequacy, efficiency and equity of higher education financing: The case of Egypt. *Prospects, 41*, 47–67.

Field, S. (2020). *Why Tunisia is suffering from brain drain*. https://www.fekr-magazine.com/articles/tunisian-brain-drain. Accessed 6 December 2021.

Fryer, Landis G., & Tavis D. Jules. (2013). Policy spaces and educational development in the Islamic Maghreb region: Higher education in Tunisia. *International Perspectives on Education and Society, 21*, 401–425.

Gdoura W. (2019). Higher education systems and institutions, Tunisia. In P. Teixeira & J. Shin (Eds.), *Encyclopedia of international higher education systems and institutions*. Springer.

Gertler, M. (2012). *Universities and cities: An intimate economic relationship*. https://www.uni-frankfurt.de/41630446. Accessed 6 December 2021.

Goddard, J., & Vallance, P. (2013). *The university and the city* (1st ed.). Routledge.

Griffiths, M., Berda, Y., Joronen, M., & Kilani, L. (2022). Israel's international mobilities regime: Visa restrictions for educators and medics in Palestine. *Territory, Politics, Governance*, 1–19.

Guessom, N. (2006). *Online learning in the arab world*. https://elearnmag.acm.org/archive.cfm?aid=1190058. Accessed 24 May 2022.

Halabi, R. (2022). *Palestinian students in an Israeli-hebrew university: Obstacles and challenges*. Higher Education.
Hoyle, C. (2015, October 28). *Palestinian students on the frontline amid daily violence*. https://www.middleeasteye.net/fr/news/palestinian-students-frontline-amid-daily-violence-535471801. Accessed 28 October 2022.
Klump, R., & Bickl, M. (2012). *The university and the city*. The President of Goethe University, Frankfurt. https://www.uni-frankfurt.de/41630446. Accessed 6 December 2021.
Krafft, C., & Alawode, H. (2018). Inequality of opportunity in higher education in the Middle East and North Africa. *International Journal of Educational Development, 62*, 234–244.
Lefebvre, H. (2003). *The urban revolution*. University of Minnesota Press.
Magnusson, W. (2011). *Politics of urbanism: Seeing like a city*. Routledge.
Masri, S. (2020). *Why Tunisia's once superior education system needs to reform again*. https://www.al-fanarmedia.org/2020/04/why-tunisias-once-superior-education-system-needs-to-reform-again/. Accessed 12 September 2021.
Mazawi, A. E. (2005). Contrasting perspectives on higher education in the Arab states. In J. C. Smart (Ed.), *Higher education: Handbook of theory and research* (Vol. 20, pp. 133–190). Springer.
Milovanovitch, M. (2014). Trust and institutional corruption: The case of education in Tunisia (Edmond J. Safra Working Papers No. 44). Harvard University, Edmond J. Safra Center for Ethics.
Ministry of Higher Education and Scientific Research. (2022). *Statistics/Public organization*. http://mohesr.gov.eg/ar-eg/Pages/StatisticsDetails.aspx?folder=/ar-eg/DocLib/2021-2020/الجامعات%20الحكومية. Accessed 24 May 2022.
MoHE, (2020). *Higher education and scientific research in numbers*. http://www.mes.tn/page.php?code_menu=13. Accessed 12 December 2021.
MoHE. (2021). *Institutions of higher education*. http://www.aqac.mohe.gov.ps/heisLinks. Accessed 12 December 2021.
Nfham. (2022). *Educational lessons for primary, middle and high school*. https://www.nafham.com/egypt. Accessed 24 May 2022.
Nussbaum, M. C. (2010). *Not for profit: Why democracy needs the humanities*. Princeton University Press.
OECD. (2019). *Gross domestic spending on R&D*. https://data.oecd.org/rd/gross-domestic-spending-on-r-d.htm.
Perry, D., & Wiewel, W. (2015). *Global universities and urban development*. Routledge.

Plackett, B. (2019). *False research results-a global problem that includes the Arab world—al-fanar media*. https://al-fanarmedia.org/2019/07/false-research-results-a-global-problem-that-includes-the-arab-world/. Accessed 24 May 2022.

Pyle, A. (1999). *Key philosophers in conversation* (1st ed.). Routledge.

Rabbat, N. (1996). Al-Azhar Mosque: An architectural chronicle of Cairo's history. *Muqarnas, 13*, 45.

Reid, D. M. (2002). *Cairo University and the making of Modern Egypt*. Cambridge University Press.

Sadeq, K. (2012). *Nafham: A learning management platform to enhance education in Egypt*. https://www.wamda.com/2012/04/nafham-a-learning-management-platform-to-enhance-education-in-egypt. Accessed 24 May 2022.

Sawahel, W. (2021). *Why Egypt's higher education sector continues to improve*. https://www.universityworldnews.com/post.php?story=20210602184615600. Accessed 12 December 2021.

Schafran, A. (2015). The future of the urban academy. *City, 19*, 303–305.

Schofer, E., & Meyer, J. (2005). The worldwide expansion of higher education in the twentieth century. *American Sociological Review, 70*, 898–920.

Schwab, K. (2018). *The global competitiveness report 2018*. World Economic Forum. http://www3.weforum.org/docs/GCR2018/05FullReport/TheGlobalCompetitivenessReport2018.pdf. Accessed 26 December 2021.

Susser, A. (2021). *Still standing, but Shaky: Jordan at 100*. https://fathomjournal.org/still-standing-but-shaky-jordan-at-100/. Accessed 26 December 2021.

Tadamun. (2018). *Inequality of opportunity in Cairo: Space, higher education, and unemployment*. http://www.tadamun.co/inequality-opportunity-cairo-space-higher-educationunemployment/?lang=en#.YciJgi2cY1I. Accessed 26 December 2021.

UNESCO. (2017). *School enrollment tertiary (% of Gross)*. https://data.worldbank.org/indicator/SE.TER.ENRR. Accessed 26 December 2021.

UNESCO. (2018a). *Research and development expenditure (% of GDP)*. https://data.worldbank.org/indicator/GB.XPD.RSDV.GD.ZS?locations=IL. Accessed 26 December 2021.

UNESCO. (2018b). *Researchers in R&D (per million people)*. https://data.worldbank.org/indicator/SP.POP.SCIE.RD.P6?end=2021&locations=IL-JO-TN-EG&start=2003. Accessed 26 December 2021.

Université de Tunis El Manar. (2022). *Université de Tunis El Manar*. http://www.utm.rnu.tn/utm/fr/. Accessed 20 February 2022.

University of Jordan. (2022). *About the University of Jordan*. http://ju.edu.jo/Pages/AboutUJ.aspx. Accessed 25 October 2022.

Walker, M. (2003). Framing social justice in education: What does the "capabilities" approach offer? *British Journal of Educational Studies, 51*(2), 168–187.

Walker, M. (2006). Towards a capability-based theory of social justice for education policy-making. *Journal of Education Policy, 21*(2), 163–185.
Wiewel, W., & Perry, D. (2015). *Global universities and urban development*. Routledge.
Wilkens, K. (2011). *Higher Education reform in the Arab world. In The Brookings project on U.S. relations with the Islamic world, 2011* (U.S.-Islamic World Forum Papers). Brookings.

CHAPTER 6

Urban Youth and Activism

Introduction

> We performed a play about a mental health institution, the patients in the institution are fine, but the doctor is the sick one and this was a representation of the Arab world and its leaders. We as Tunisians, Palestinians, Egyptians; we want to rise, but the leaders are stuck.[1]

As we have seen in the previous chapters the political, economic and social conditions in Arab cities have affected the youth segment negatively, at varying levels of severity, based on gender, social class and ethnicity. The cities we studied are not offering their citizens/residents with a strong public pre-school education system, also public-school education is not sufficient in building their capabilities and is rife with inequalities, similarily, tertiary education is overrepresented with youth from wealthier segments of society. Economic structures are also making it harder for youth to acquire stable employment or create their own entrepreneurial projects or businesses, thus making it harder for them to transition from their family home. The political and economic structures

[1] Interviewee Basem (Pseudonym), 24, Muslim, Male, Performer, Tunis.

R. A. Nuseibeh, *Urban Youth Unemployment, Marginalization and Politics in MENA*, Middle East Today,
https://doi.org/10.1007/978-3-031-15301-3_6

in our four cities are preventing youth from full independence and democratic participation. The depilated urban living conditions, economic failures and deficiencies in education and health care, in combination with police brutality, corruption and authoritarian control of urban spaces, together, severely limit civic participation of urban youth (Ginwrite et al., 2005). Arab youth access to formal power is marginal, but they are not passive actors in their own communities. They have shown the power to revolt against structural injustices, as was reflected in the revolts that shook the streets of Arab cities in 2011, the two Palestinian *Intifadas*[2] and the continuous protests taking place on Palestinian streets.

Various grassroot organizations, civil-based institutions and groups, have managed to organize the fight for city residents' rights; some groups advocate and act alone, while others have managed to build bridges and construct alliances with other groups within the city, to envision alternative possibilities and fight for a more just city, that serves all (Miller & Nicholls, 2013). These alliances can be very fruitful in helping marginalized groups navigate the complex restrictive urban policies, by pooling together their collective knowledge, experiences and strengths (ibid.). Digital activism and the wide reach of media, have also enabled civil groups to support each other across borders, as witnessed in the Middle East. Every time the Palestinian streets get shaken by civil protests and unrest, cities in the region and even around the world start demonstrating in solidarity (Al Jazeera, 2021). Local urban unrest can also trigger revolts around similar political and economic injustices, as we have seen when the revolts in Tunisia inspired revolts around the Arab world.

The City, Exclusion and Social Movements

Cities are central sites of social change; on the one hand—as we've seen in the previous chapters—they can be sites of injustice and inequality at higher rates than the national level, and on the other hand, they can be sites of mobilization to fight the political and economic structures dictating life in the city (Espinosa, 2018). Neoliberal and hegemonic political policies have resulted in deep urban inequalities, dispossession and suffering among large segments of impoverished city dwellers. When the logic of profit dictates the direction of urban development, fewer

[2] Intifada: Arabic word for shake off, and it represents the Palestinian uprisings of 1987 and 2000.

investments will be made in social infrastructure, and more incentives will be given to investors at the expense of the environment and the urban community. This can be witnessed acutely in developing cities struggling to catch up with the "First World's" rate of development, and in their bid to attract big market players, they are sacrificing their residents and the environment by offering cheap land and labour (Domaradzka, 2018). This approach results in discriminatory urban policies and systematic displacement of residents, based on their social class and ethnicity, which is then enacted through bureaucratization processes. However, this systematic displacement of urban residents from the center, and the unjust policies that reproduce their poverty and deprivation, can serve to catalyze them to assert their power and find their space within their cities (Miller & Nicholls, 2013). This means that cities, as much as they are sites of struggles and inequality, can also be centers of revolution, where people can mobilize, to contest social, spatial, economic and political injustices. Uitermark et al. (2012) argue that cities breed contention, because unjust urban policies can produce a variety of grievances among city residents, hence offering an opportunity for various marginalized groups to formulate alliances and mobilize to resist and change urban realities. The city itself as a center of concentrated power and privilege, can also present a point of attack. Uitermark et al. (2012) argue that the task of social movements analysts is to then understand how contention is occurring; when and where it's erupting; and what political ideology connects this contention to other actors beyond the local level, and what turns this mobilization into a regional and even a global movement.

Throughout history urban movements have included attempts by city residents to take control of their urban space, and to be included in shaping their space. When there is an absence of effective institutionalized formal mechanisms for them to challenge injustices, or claim their space in the city, residents have resorted to resistance and urban activism. Zemni (2017) suggests that the success of national revolutions in Arab countries is intertwined with urban revolutions, and people's ability to assert their "right to the city". Urban activism can take several forms; such as trade unionism, grassroot and neighborhood initiatives, social Islamism and non-governmental organization activism, and urban mass protest (Bayat, 2002; Domaradzka, 2018) as well as creative resistance through art and culture jamming,[3] which were major tools used in the 2011 Arab uprisings.

The Role of Unions in Resistance

An important form of urban activism in the context of Arab countries is trade unionism. Trade unions emerged in Arab countries in the context of colonial domination, and so they were fighting both class and national hegemonies (Bayat, 2002). However, after independence, most trade union organizations in Arab countries were integrated into state structures or the ruling parties, and their leadership co-opted, rendering union activism ineffective, with many of them becoming part of ruling parties and the state bureaucracies (Bayat, 2002; Hibou, 2011). A key example, is that of the Tunisian General Labour Union (*Union Générale Tunisienne du* Travail, UGTT). In 1952, as the nationalist Néo-Destour party was repressed by the French colonial authorities, the UGTT became the major political power, working for independence with around 150,000 members (Today around 1 million members) (Netterstrøm, 2016). Aware of its political weight, subsequent post-colonial authorities sought to gain control of the union, starting with Bourguiba's government in 1956, then followed by Ben Ali's government (ibid.). Hibou (2011: 124) described the UGTT as a "a pure and simple appendage to power." Bellin (2012: 139) in her analysis on the robustness of authoritarianism, also argued that labour unions in Arab countries were "empty shells." She also argues that the weakness of labor unions and civil society institutions, is detrimental to developing countervailing power, that can force the state to be accountable to people's needs. Another issue with the weakness of labour union activism is the fact that most labor is now informal, precarious and not organized, and there is no protection for workers, and most workers are on temporary contracts (Bayat, 2002).

The revolution of 2011 changed the role of labour unions, especially in Tunisia. Netterstrøm (2016) tells us how the UGTT managed to survive the authoritarian rule of Ben Ali by utilizing both resistance to the regime, and cooperation with the regime, which made it possible for it to survive, whilst also playing a role in the revolution. During Ben Ali's regime, corruption seeped through to the leadership of the UGTT,

[3] A term coined by Dan Joyce in 1984.

leading to complicity with the regime. However, the subjugation of the union was not uniform, as some regional unions led by militants with socialist backgrounds were able to hold some level of autonomy (Zemni, 2017). This dual complicity and resistance by several members within the union, made it possible for the UGTT to survive the dictatorship, and to have a decisive role in the revolution (Netterstrøm, 2016). During the revolution, their offices served as meeting points, banner distribution centres, and safe refuge for demonstrators; the regime was reluctant to raid their offices, because of the cooperation with higher level members of the union (ibid.). In their congress meeting in 2011, the UGTT managed to install a new leadership made up of left wing and anti-Ben Ali members (ibid.). The UGTT continued to play a major role after the revolution, by drafting the road map for the constitution, in collaboration with different political parties and civil society organizations. The constitution passed by a majority, in the National Constituent Assembly, and earned them a Nobel Peace Prize (ibid.). Most recently, the UGTT has organized several strikes, one of them taking place in 2018, with large participation from civil servants protesting the IMF's recommendation to the government to freeze public sector wages and reform companies to address the public deficit (Ben Yahmed & Yerkes, 2018).

The Egyptian Trade Union Federation (ETUF) was established in 1957. Today it is the largest trade union center in the Arab world, with 7 million members (ETUF, 2022). The ETUF is a legally prescribed monopoly, under which all unions are required to belong to (Blackburn, 2018). The union leadership was always integrated into the ruling party in Egypt, shedding doubt on how much the union really worked in its labourers interests. The union was also never democratic, and not everyone was permitted to compete in the electoral process. The most recent presidential elections witnessed the installation of a Sisi supporter, and the election was described by Human Rights Watch as 'farcical' (Blackburn, 2018: 12). With that said, Egyptian workers have managed to organize themselves outside the framework of the union to push for change. For example, the strikes of more than 24,000 textile workers in Mahllah in 2006 are said to have been a major milestone towards the Egyptian uprising of 2011 (Bayat, 2015). After the uprising in 2011, the Egyptian Federation of Independent Trade Unions, and other independent trade unions, were established to fight for workers rights, in defiance to the ETUF monopoly over trade unionism. However, the newly formed unions continue to suffer from organizational weaknesses, including lack of transparency, limited funding and internal leadership struggles, impacting on their ability to serve the workers (El Sharkawy & El Agati, 2021)

The union in Jordan also has a similar history of undemocratic electoral processes and collaboration with ruling parties. The General Federation of Jordanian Trade Unions (GFJTU) was founded in 1954, and has a membership of 135,000 workers today. It is the national trade union center in Jordan and has 17 affiliated unions (ITUC, 2019). The union has been described as a "semi-governmental institution," due to its financial dependence on the government (Jordan Labor Watch, 2012: 7). The affiliated unions followed unified and non-democratic by laws, that allowed for a small number of people to dominate the leadership of trade unions for extended periods of time, leading to the absence of internal union elections (ibid.). With the wave of uprisings engulfing Arab countries, which called for justice and democratic practices, 7000 Jordanian workers, represented by nine independent unions, held the founding congress of the Federation of Independent Trade Unions of Jordan (FITU-J) in 2013 (Connell, 2013). This movement was pushing for more independent trade unionism, and for greater democratic practices, political freedoms and improved economic conditions and justice (ibid.). The new FOTUJ has been attacked by the GFJTU, which called for its banning. Since 1976, no new trade union has been allowed to form in Jordan, and this drastically harms the large numbers of migrant workers in Jordan, who are not allowed to form a union (Connell, 2019). This limits their ability to fight for better working conditions. Jordan labor laws permit only one union per sector, and there are unions in only 17 sectors set by the government, further limiting union activity and independence (Connell, 2019). A recent amendment to the law in 2019 gave the labor ministry the power to dissolve unions, and impose fines on those that continue activities in the dissolved unions (ibid.). Both Jordan and Egypt have a long history in breaching the International Labor Organization conventions.

The seeds of Palestinian trade unions started at the end of the British mandate period. These unions were led by the communist party and were mainly representing a minority of urban industrial workers. Later, under Jordanian rule in 1953, labor law codified union activity and unions expanded in Palestine, despite the banning of the communist party and the shift of the control of the unions to the Jordanian government (ibid.). The Palestinian Federation of Trade Unions PGFTU, that is present today, is derived from the Jordanian Trade Union parent organization in 1969. Since its inception, the union was always political and splintered along faction lines. Therefore, since the beginning of the Israeli Occupation of the West Bank, union leaders were targeted and imprisoned. The effects of Israeli Occupation—between land confiscation, expansions of Israeli settlements on Palestinian land and movement restrictions, have all had their toll on Palestinian livelihoods, leading many Palestinians to turn into wage laborers in the Israeli market and Israeli settlements. These workers suffered exploitation as a result of both capitalism and Israeli colonialism, and they were neither protected by the PGFTU nor the Israeli *Histadrut* (The General Organizations of Workers in Israel). Since the establishment of the Palestinian Authority, top union leaders were co-opted by the PA, and the PGFTU has capitulated to the demands of the authority. It is not democratic, does not hold elections, and does not have the capacity or the political will to fight for Palestinian workers.

Forms of Resistance and Activism

Neighborhood or community-based collective activism is a form of resistance in urban spaces. During the first few weeks of the Tunisian revolution, while union militants organized non-violent sit-ins during the day, neighborhood youth organized night riots against the authorities (Zemni, 2017). Another example of neighborhood activism and resistance, includes the mobilization of people in East Jerusalem's neighborhood of Sheikh Jarrah, who were risking forced evictions by the Israeli municipality. They used social media, demonstrations and sit-ins and managed to mobilize Palestinian NGOs, politicians, artists and academics, as well as Israeli left-wing organizations to join their fight to keep their houses. Another example by Bayat (2002) of urban social movements, includes the campaign of the low-income community of Ezbat Mekawy in Cairo, against industrial pollution in the area. They also engaged the media and lobbied politicians to protect their community.

Another form of activism in the Arab world can be seen in Islamist movements and their social development agendas. In light of the withdrawal of the state from providing welfare services, Islamic non-governmental organizations have flourished and provided communities with social services that the state was not providing, such as health care and financial aid (Bayat, 2002). Although Islamism may be considered a social movement, it is not particularly an urban movement that cares for the rights of the subaltern; their ideology is mostly derived from applying Sharia law and applying Islamic teachings (Bayat, 2002). Similar to Islamic NGOs, there are various secular and international NGOs and aid organizations, that work on supporting urban marginalized groups in Arab cities. They also work on raising awareness and lobbying to bring attention to various issues, such as gender inequality or the environment. However, although the Arab world is filled with thousands of NGOs, the scope of their work remains small, unsustainable and not organized.

The strongest and most pervasive form of urban social activism taking place in Arab countries is what Bayat (2002: 19) refers to as the "quiet encroachment of the ordinary," which he defines as the silent pervasive advancement on spaces in the city, through unlawfully acquiring land, building homes, and getting urban services, jobs or business opportunities in a quiet, prolonged fashion. The idea is that if the current structure and bureaucratic systems that rule the city are exclusionary, then people will start creating their own law and their own reality. The problem with this kind of "quiet encroachment" and "social non-movements", as Bayat (2002, 2015: 34) refers to it, is that it's chaotic, and it doesn't have an ideological leadership or organized structure. Bayat (2002) explains that within the authoritarian political structures in Arab countries, "quiet encroachment" is the strongest form of urban social activism. He believes that the scope of grassroot activism, social Islam and NGOs are limited, and have failed to improve living conditions for people. This quiet action is taken over long periods of time, without any clear leadership or ideology, just people in the act of claiming services and spaces in the city, such as the urban poor tapping electricity from municipal power poles, or connecting to water pipes, or claiming spaces on the street to do business informally, without paying taxes to the municipality, or building without getting a municipal license (Bayat, 2002). Allegra et al. (2013: 1677), in their analysis of protests in Arab cities, have also noted that protests are not the only and most significant phenomena occurring in cities during contentious times; cities should be viewed as "points of access" or "privileged sights", where political and ideological struggles can be analyzed with their various forms and meanings to the people experiencing them.

Repressive municipal authorities work hard on crushing this kind of activism. In 2020 alone, Egyptian authorities referred around 6,000 building violations to military prosecution and demolished more than 21,000 unlicensed constructions around the country (Ahram, 2020).

In East Jerusalem, in particular, the Israeli municipality is engaged in a relentless campaign of house demolitions and heavy fines, in addition to incarceration of workers without permits, coming from the West Bank, and confiscating and fining street vendors without permits. According to Betselem (2021), between 2004 and 2020, Israel has demolished 1,097 Palestinian houses, and 458 non-residential units built without permits. Since its establishment, it is estimated that Israel has demolished more than 131,000 Palestinian homes (ICAHD, 2022). East Jerusalem and the old city with its religious sites were a hub for Palestinians coming for prayer, and also street vendors selling agricultural produce from all over Palestine. The creation of the Israeli Separation Barrier, and the stifling of Palestinian agriculture and small framers' businesses, have rendered street vending in East Jerusalem very limited.

In Cairo, municipal authorities are also attempting to assert state power over public spaces. Following the fall of the Mubarak regime, successive governments have engaged in prohibiting public assembly, informal economic activity and creative artistic activity from *Tahirir* square, which is the symbol of urban centrality in Cairo (Abdelrahman, 2013). The government engaged in several projects aimed at evacuating central districts from poor people, such as creating new urban markets in Cairo's outskirts (Nagati & Stryker, 2013), or moving people from central informal settlements to new cities on the outskirts. These projects have been a consistent failure (Sims, 2014). The same struggles can be seen in the market in downtown Tunis, with hundreds of street vendors working without municipal permits. They get subjected to daily raids by the police, which include confiscation of goods and detention, but despite these risks, for many people this is the only option for survival (Dreisbach & Smadhi, 2015). Corruption among Egyptian and Tunisian municipal authorities has also increased people's suffering and resentment.

> In the southern Giza district of Haram, as soon as people learn that municipal patrols are about to enter the district, they rush to close their coffee shops, while street vendors attempt to dodge the patrols.... Bribes are seen as the only other way around the patrols' inflated enforcement efforts. (Rabie, 2019)

Municipal employees in Cairo are also constantly being bribed to issue routine construction, water and electricity licenses for new small business-like restaurants and shops (ibid.). Harassment of street vendors by municipal authorities is prevalent in Arab cities. It was actually the self-immolation of the Tunisian street vendor Bouazizi that sparked the uprisings of 2011 in the Arab world, because people identified with him. As De Soto (2011) says that the 180 million Arabs who work in and around the informal markets in the MENA region have identified with Bouazizi-a street vendor, who works in the informal sector and had his merchandise confiscated by municipal authorities, under the claim that he didn't have the formal papers needed. De Soto says that like 50% of all working Arabs, Bouazizi failed to find employment in the formal sector, and was trying to do business on the margins of the law, in an informal economy, and he died trying to do business while hassled by corrupt municipal authorities. Quoting Bouazizi's brother, De Soto writes "the poor also have the right to buy and sell." It is this exclusion from participating in the "free market" that pushes people to dissent.

Bush (2004) actually contradicts this terminology in describing the exploitation of the poor in the neoliberal economy in the MENA region. He argues that poverty is not resulting from excluding people, rather it is the result of an economic environment that is built on exploiting their labor. The term "free market" is deceptive, as Bogaert (2013) explains, what occurred in Arab countries is an economic and political shift from general social progress, to an indirect control of the economy through the redistribution of resources, to domestic and foreign elites. This is evidenced in the high rise real estate projects referred to as the "Dubai effect" (Barthel, 2010: 133), where profit is accumulated in the hands of the few, at the expense of the majority impoverished residents of the city. It is this exploitation, and the deprivation of people from leading the life that they have the desire to lead, added to that their increased awareness of this injustice—due to exposure to the internet and the media—that pushed for the mobilization of people in the Arab world. As De Soto (2011) says, if we learned anything from Marx, it's that when the powerless become conscious that they share a common suffering, they can transform into a revolutionary class.

Claiming the City Center

The desperate sporadic struggles to gain rights in the city, and the different forms of mobilizations that we discussed, can turn into organized collective resistance, when the opportunity becomes available. Turning those daily struggles and injustices into an actual organized urban mass protest takes time. Resistance against the dissolution of the social welfare policies of the 50's and 60's in Arab cities, and the austerity measures that accompanied the neoliberal era of the 70's, started towards the end of the 70's and 80's,[4] when the Arab region began to witness "urban riots" and "bread riots" (Bogaert, 2013: 225). Egypt experienced numerous labor strikes and violent clashes ahead of and leading up to the 2011 uprising. There are even arguments that the demonstrations of Mahalla 2006 in Egypt, and Gafsa 2008 in Tunisia, were the forerunners for the uprisings of 2011 (ibid.). Bayat (2015) explains that mass urban protests can happen when the state slips into crises, or some large social movement gains momentum and starts to organize dispersed struggles around the city. In the case of Cairo, politics began to change in the early 2000's with the formation of the Popular Committee for Solidarity with the Palestinian and Iraqi people, "the Kefaya" democracy movement, "the April 6th Youth" Movement and "We are All Khalid Said" Movement (Bayat, 2015; Lim, 2012).

Kefaya (meaning enough in Arabic) is the slogan of the Egyptian Movement for Change (*el-Haraka el-Masreyya men agl el-Taghyeer*). It was founded in 2004 by Egyptian academics, from various ideological backgrounds; it was not established as a political movement, rather as a coalition of political forces who shared a call to end Mubarak's rule (Carnegie Endowment for International Peace, 2010). The movement organized protests starting in downtown Cairo, outside the High Court, followed by protests on university campuses and in the famous Tahrir square (ibid.). It's in the center of this urban space that the first demonstrations started flourishing and mobilizing groups for change. The powerful slogan Kefaya (enough), spoke to urban dwellers fed up with injustices, authoritarianism, lack of democracy and equal opportunities. These demonstrations then spread throughout the country. The scope of Kefaya's work in Egypt remained small, and they never managed to become a political power. By 2006, the movement almost disintegrated and disappeared from the public sphere. The movement faced several challenges; on the internal level, it had friction with the Islamic groups within the movement, that withdrew from the coalition, hence weakening the movement's momentum (Carnegie Endowment for International Peace, 2010), and externally, they were being brutally attacked by

[4] Egypt (1977), Morocco (1981, 1984), Tunisia (1984), Sudan (1985) Algeria (1988) and Jordan (1989).

state media, incarcerated, intimidated, and sexually harassed and assaulted by the state regime. Although the Kefaya movement has disappeared from the political arena, it managed to inspire online youth activism, particularly on Facebook and Twitter, starting in 2008 (Lim, 2012) and later on Instagram as well. It managed to ignite a spark in the country, and make a crack in the authoritarian control of dissent. It has also managed to use information technology effectively to spread the message among the Egyptian public; by using text messages, online advertisements and online caricatures and cartoons to disseminate their message and advertise their rallies. Street protests were organized and advertised digitally- 54 out of 70 recorded street protests from 2004 to 2011 in Egypt involved online activism (Lim, 2012). The use of social media is more prevalent in urban spaces in particular, with higher access to internet. In Egypt, leading up to the 2011 revolution, on the national level, access to the internet was 30%, while in Cairo the level was more than double that, at over 64% (Lim, 2012). They also managed to fuel public sympathy by documenting and digitally disseminating the abuse of security forces. Social media has played and still plays a major role in recruiting individuals, mobilizing them, lobbying and adding pressure on politicians and powerful entities. Social media has become a battleground, a space for the marginalized to express grievances, and bring global attention to their cause, raise awareness and disrupt authoritarian regimes. Another leading social movement group that has managed to utilize the internet to spread its agenda and rally people, is the "April 6 Youth Movement". In March 2008, a group of young tech-savvy Egyptian activists launched a Facebook page in support of the famous textile worker strike in Mahalla city in 2006, protesting labour conditions and wages; within a few weeks the page managed to gather over tens of thousands of followers. Although group leaders were subject to harassment and interrogation by state security, they still managed to keep their movement going and had a leading role in the revolts of 2011. Inspired by the Tunisian revolution, the "April 6 Youth Movement", together with the "We are All Khaled Said" movement, played a crucial role in the 2011 revolts in Egypt (Carnegie Endowment for International Peace, 2010b). We are All Khalid Said Facebook page was created to rally Egyptians against police brutality and corruption. Khalid Said was a 28 year old who was taken from an internet café and beaten to death by the police, allegedly targeted because he was in possession of video footage showing corrupt police officers sharing drugs confiscated in a drug bust (Chick, 2010 cited in Lim, 2012). Graphic images of the brutalized body were circulated and the Facebook page managed to gather 5 million subscribers. The two movements have used both online and physical tactics to rally for their famous *Tahrir* Square demonstration. They distributed flyers, used Taxi drivers and coffee shop owners to spread word of mouth about the demonstration, they also mobilized soccer fans, in addition to their online activism and rallying (Lim, 2012). The poor and disenfranchised, some of who did not have access to internet were also having their own struggles prior to the 2011 events. Bayat (2015) tells us that several urban protests took place prior to the outbreak of the 2011 revolt, undermining police authority, empowering urban residents and paving the way for the revolution. He tells us of several strikes and demonstrations taking place four years leading up to the 2011 uprising. In addition to the Mahalla strikes, that we mentioned before, there were several strikes and demonstrations, including the strike of the 150,000 self-employed garbage collectors in Cairo, protesting the government slaughter of swine in 2009, and demonstrations in informal settlements against demolitions, lack of water access and inequalities in neighborhoods around Cairo, where the police proved unable to contain the protests within enclosed spaces (Bayat, 2015). It is within this political and economic atmosphere of gentrification, eviction of poor communities, lack of services and jobs, as well as people's increased awareness, and the new online strategies that the state apparatuses were not ready for; all these factors combined allowed for the mobilization of the masses for the January 2011 uprising.

When the protests in Egypt started, they were never about overthrowing the regime, and more about claiming urban rights. The Tunisian revolts actually refocused Egyptian oppositional movements, and gave them hope that change was possible and that they have the power to implement it (Lim, 2012). Even in Tunisia, the initial protests did not aim to remove Ben Ali from power, but as the protest grew in size, political opposition figures seized the opportunity to call for a change in regime (Ben Yahmed & Yerkes, 2018). What raised the people to protest, was the feeling of injustice and exclusion. The sense of injustice differed in the Tunisian context across the country, from lack of access to land in rural areas, to the perceived sense of theft of resources in the mining regions, to the grievances against corruption and nepotism that the middle class suffered from (Zemeni, 2017). They all managed to go under one umbrella and ask for a change in the regime. The urban played a major role in this, with major public spaces gaining a new symbolic meaning (Zemini, 2017).

Inspired by the revolts in Tunisia and Egypt, Jordanians started protesting the austerity measures followed by the government, calling for better wages, restoring food and fuel subsidies, and demanding an end to the experimentation with IMF mandated market reform (Schwedler, 2018). Protests started in Dhiban, a town located south of Amman in the Madaba Governorate. Protesters' anger stemmed from loss of jobs at a nearby phosphate factory after its privatization, as well as from the devaluation of their land, after the creation of a dam that channeled water from their town to Amman (Pelham, 2011). To quell dissent, the regime followed the stick and carrot policy. To pacify the protesters in 2010, the government spent an emergency $550 million economic package, to raise government salaries and pensions, and partially reinstated food and energy subsidies (Tobin, 2012). However, and at a later stage, as the country's deficit increased, it cut both fuel and food subsidies to secure an IMF loan. That loan was intended to rein in a budget deficit that threatened the country's fiscal and monetary stability, but it came with a high price and crippling austerity measures. These measures hit people with the lowest socio-economic status the most, adding to the wealth gap and inequality.

Demonstrations continued, but they never reached the scale of demonstrations in Tunisia and Egypt. They were mainly held by East Bankers. Palestinians who constitute 60% of the Jordanian population abstained or had limited participation in these demonstrations. Pelham (2011) argues that the reasons for the silence of West Bankers in these protests is their trauma from 1970 crackdown. Palestinians usually champion foreign issues like Israel's wars in Lebanon and Gaza, rather than interfering in domestic politics in Jordan. They also do not have the tribal support that East Bankers have. Jordan has also strict laws prohibiting freedom of speech. Jordanian law criminalizes speech deemed critical of the king, foreign countries, government officials and institutions, and Islam and Christianity (Roth, 2021).

Protests in Jordan did not manage to gain enough momentum and topple the ruling regime. The major opposition elements that led the mobilization against the regime included the typical opposition group, which is the Muslim Brotherhood, and new actors such as the group of military retirees, who formed a committee in 2007 to represent 150,000 military retirees, the "Group of 36"—which is 36 personalities representing prominent Jordanian tribes resentful of the political and economic climate in Jordan, and the "Jordanian 24 March Youth Movement", which is a youth movement that capitalized the use of social media to mobilize people demanding political reform (Al Shalabi, 2011). Another youth movement was *Thabahtoona* (you have slaughtered us), a movement formulated by a group of students to fight the rising costs of higher education and to voice the concerns and the hardships faced by tertiary education students in Jordan (Cantini, 2012). In 2012, another protest movement developed called *Al Hirak* (The Movement), its participants were Transjordanian Youth. Several protests took place during that time, mostly organized by Islamic Action Front (IAF), which is the political wing of the Muslim Brotherhood in Jordan, mostly after Friday prayers. However, the protests never managed to mobilize large enough numbers of people, like the protests in Cairo and Tunis. The largest protest occurring in Amman, during that time, managed to attract around 10,000 people (Tobin, 2012). The protests were calling for political reform, increasing the power of the legislature vis-à-vis the executive, firing the prime minister, amending the constitution and amending election laws to stop gerrymandering (Harris, 2015). There were, in fact, widespread allegations of election tampering, and a clear lack of independence of municipal authorities, and parliament was incapable of generating legislations (Peters & Moore, 2009). In addition to demands for political reforms, people were also protesting rising prices and high unemployment. The regime tried to pacify protesters, by doing some cosmetic reforms to its electoral laws. While the regime managed to contain a full en masse protest that could have lead to its toppling, it is still dealing with bouts of demonstrations. As Susser (2021) says Jordan at 100, "still standing, but shaky."

The current cohort of Palestinian youth is sometimes referred to as the "Oslo generation"—people who are born after the 1993 Oslo accords and that are described as the generation that is politically alienated, fragmented and economically marginalized (Dwonch, 2019: 1). Palestinian youth are fragmented geographically, between East Jerusalem, West Bank, Gaza, Israeli territory and international diaspora, but they are united in the struggle for the Palestinian cause, and they share that on social media platforms that cut across geographic segregation. Dwonch (2019) has studied Palestinian youth online activism, which she dates back to 2011 and believes that it was also inspired by the Tunisian and Egyptian uprisings. It started with online calls to end the *Fatah* and *Hamas* faction divisions "The people want the end of the division," followed by "Palestinian Freedom Rides," exposing the expansion of the Israeli settlements and military checkpoints (Dwonch, 2019: 2). Youth disillusionment with established Palestinian political parties, and their frustrations with neoliberal Palestinian economic policies in the West Bank and complicity with Israeli Occupation policies and the colonial expansions over Palestinian spaces, have increased online youth activism, both against the Palestinian Authority and Israel. Politically unaffiliated youth were participating in various demonstrations, whether in the West Bank, protesting the murder of Nizar Banat, an activist and critic of the Palestinian Authority (Amnesty International, 2022) or the protests of "*Habat Al Karameh*" (the uprising of dignity), where Palestinian Israelis and Palestinians in the West Bank went into the streets to protest the Israeli Occupation (Wated, 2022). Through their social media accounts, the Palestinian Al-Kurd siblings have managed to create global awareness on the ethnic evictions that the Israeli Jerusalem municipality is committing against Palestinian residents of Shiekh Jarrah neighborhood in Jerusalem (Mansoor, 2021). Palestinian civil society organizations, and international and Israeli human rights organizations, have also used social media platforms to report on the human rights breaches committed by the Israeli Occupation forces.

Massive mobilization took place in 2017, when Israel decided to place electronic security gates at the entrance to Al Aqsa Mosque (Beaumont, 2017). Demonstrations shook the streets of the Old City of Jerusalem, leading Israel to retract its decision of restricting access to worshippers. Another recent mass mobilization, at the heart of the city center of Jerusalem, took place around the funeral of the murdered Palestinian American Journalist Shreen Abu Akleh, who was shot by Israeli soldiers, while covering their invasion of Jenin refugee camp. The funeral witnessed mass mobilization of mostly politically unaffiliated Palestinian mourners, who claimed the streets of the city waving Palestinian flags and reclaiming the Palestinian identity of Occupied East Jerusalem (Vohra, 2022).

Youth Art and the Center of the City

The regime, represented by municipal authorities, police and security services, has tried to limit and control people's claim to public spaces. Arab youth, on the other hand, have reclaimed those central urban spaces during the uprisings, not only by organizing protests, but also through visual art. The walls of Arab cities were painted with messages of defiance. During the uprising, artists in Egypt have relished on the new atmosphere of freedom of expression, so Egypt has witnessed a flourishing of graffiti, street performances, caricatures, visual arts, underground music and satirical videos (Amin, 2020), and similar trends were witnessed in Tunisia. Various types of creative and critical street art and graffiti were showing up on major squares and streets leading up to them. Public buildings of the parliament, law courts and government ministries, which were all symbolic central spaces that have been appropriated by the authorities and demonstrate their dominance over the masses, were also targets of street artists, in defiance of the authorities (Tripp, 2013). The protesters have reclaimed these spaces in the name of the excluded, repressed and denied masses. It was a fight over urban centrality and dominance. In Cairo, the central Mohammed Mahmoud street, just off Tahrir Square and leading to the interior ministry, has become a symbol of resistance. The walls of the street were filled with street art, chronicling the events of the uprising, depicting the dead and the wounded, while employing elements from ancient Egyptian art (Lau, 2013).

Pro-regime art and symbols were also destroyed and defaced, sometimes in an artistic manner to ridicule the regime. The regime has also used visual art, prior to the uprising, to reinforce its authority, by installing paintings, sculptures and pictures of political leaders. Actually, visual art and destruction of regime art, also started years before the 2011 uprisings. One recorded event was during the famous 2006 Mahalla textile strikes, where workers destroyed and trampled on the picture of Husni Mubarak (Tripp, 2013). The regime was very swift in controlling the distribution of photos and videos documenting this event, as it reflects cracks in its control over the people (ibid.). Palestinian street art has always been a symbol of resistance, murals painted on the Israeli Separation Barrier, which has been used as a canvas of resistance by both Palestinian and international artists including Banksy (Andrews, 2020). Even prior to the construction of the Separation Barrier, Palestinian street art has been a major symbol and tool of resistance. Naji Al Ali, the most prominent Palestinian and Arab cartoonist, depicted the suffering of the Palestinian population through his famous refugee child figure *Handala*. His fiery cartoons that criticized the relationship between the US, Israel and Arab regimes were widely spread in the Arab world (Handala, undated).

Most recently, murals were drawn on the walls of Sheikh Jarrah, to claim the space, and to represent the resistance of the forced political evictions of Palestinian families from the area. The Israeli municipality responded by whitewashing these murals (Al Haq, 2021). Same was done by the authorities in Cairo that whitewashed most of the uprising's street-art in a claim of "tidying up" the city, and renovating downtown Cairo. Nearly all of the street art that was in and around Mohamed Mahmoud Street has been whitewashed, and an iconic graffiti wall partially torn down (Amin, 2020). Today there is hardly any trace of the cultural awakening that took place during the 2011 uprisings, and the years leading up to them, with an ongoing authoritarian security crackdown on artists, journalists and human rights activists since 2013 (ibid.).

Post Arab Uprisings

Today, over a decade after the uprisings, the social and economic ailments that pushed the youth to protest are still present. The latest Arab Youth Survey (2021) shows that the youth residing in Tunisia, Egypt, Palestine and Jordan have deep concerns about the economy, the low quality of educational and health services in their countries, and their personal freedoms. Youth are also dissatisfied with their government policies, lack of political reform, high levels of corruption and lack of democracy (ibid.). A large number of youth believe that anti-government protests could take place in their country over the next year (ibid.). Sporadic protests have actually never stopped in Egypt, Tunisia, Jordan and Palestine. Jordan has witnessed several protests over gas agreements with Israel in 2014, protests over new income tax in 2018, and the march for unemployment took place in 2019; in 2021 the government cracked down on protests against its mismanagement of the COVID-19 pandemic (Susser, 2021). The Sisi government has cracked down on all forms of dissent. In 2014, the Egyptian court banned the activities of the "April 6 Youth Movement" and accused them of espionage. State media, and pro-government private media outlets in Egypt, use this day to celebrate the president and the security forces (Reporters without Borders, 2021). The Sisi government has done a major crackdown on media outlets in Egypt, with 500 websites blocked, including the website of the last major independent media outlet *Mada Masr* (ibid.). Pre-emptive government raids and rounding up of activists is a common tactic by the authorities to crack down on dissent. Human rights organizations say that there are around 60,000 political prisoners languishing in Egyptian jails (Yee, 2022).

> Another problem is political oppression; nobody is free to say whatever he wants. I was really happy at the AUC, to attend a political debate. I could say anything I wanted without worrying that someone is going to take me to jail. Honestly, freedom of speech really matters, it makes us feel that we are human beings. Whenever I'm on campus I feel safe, whether discussing internal issues or global issues; I don't have any worries discussing them, but the minute I step outside the campus, that's it, we are totally oppressed. You cannot discuss anything; you cannot even mention the name of the president of your country. This is really weird. This is really bad, and it's one of the major reasons why people emigrate.[5]

However, even with these repressive measures anti-Sisi protests are widespread, unpredictable and not only organized by the usual suspect—the Islamic Brotherhood activists, so it is harder for the regime to contain. Protests started in 2016 despite the infamous 2013 anti-protest law, that banned mass public gatherings; people were outraged at Sisi's maritime agreement ceding sovereignty over two islands in the Red Sea to Saudi Arabia (Aman, 2016). Another major eruption in protests took place in 2019, as Sisi was headed to the United States to attend the United Nations General Assembly. Protests broke out in eight cities around Egypt, with the biggest crowds gathering around Tahrir square, pictures circulating on social media were reminding people of the 2011 uprisings that ousted Hosni Mubarak, with similar chants being used in the protests. The protests broke out, as a self-exiled contractor called Mohamad Ali, who worked for the army for over a decade, accused Sisi and his aids of corruption. He called for people to go out after a football match and protest. Football matches and Friday prayers are two events that have resulted in the break out of protests, not just in Cairo, but around the Arab world. Most of those attending the protests are young people upset with the current economic situation, corruption and authoritarian rule, similar to the protests from the previous decade.

Youth in Arab cities are still targeted by the authorities and lack a sense of safety and security. In Egypt and Jordan, young bloggers, artists and activists, still face harassment, imprisonment and torture at the hands of

[5] Interviewee Salim (Pseudonym),19, Muslim, Male, Political Science Student, Cairo.

authorities. Palestinian youth have suffered at the hands of both the Israeli and the Palestinian authorities as well, whether protesting the Israeli Occupation or protesting the corruption of the Palestinian Authority. Young women are also specifically targeted; women who have been leading mobilization campaigns and participating in protests have suffered from sexual harassment and smearing at the hands of the authorities.

In their study, "Power 2 Youth", Calder et al. (2017: 14) argue that policymaking within Arab countries tends to focus on restricting youth spatially to areas where they are invisible, or excluding them from spaces where their "presence is not required for business." They say that little effort is made to include youth in decisions about urban or spatial planning, and youth interests do not feature in planning priorities. Even though youth movements have managed to capture these central urban public spaces with their protest, with their art and their creativity, today, in Cairo, Amman and Jerusalem, authoritarian control is very much felt by youth. Tunis, on the other hand, is going through its fragile transition phase. Protests broke out in Tunis when the President suspended parliament, sacked the prime minister and expanded his legislative and executive powers, in a move that many in Tunisia considered a coup of the government. Protests over the economic crisis and corruption have also fueled Tunisian streets since the revolution. The COVID-19 pandemic has increased unemployment in Tunisia from 15.3% to 17.4%, leading to massive frustrations among Tunisians. The combination of the political unrest and the economic crisis has led to an increase in protests—there were 3,865 cases of civil unrest in the first quarter of 2021 alone (Abdselm, 2021).

Conclusion

Cities can be sites of both inequality and marginalization, as well as central arenas for the fight for equal rights. The logic of profit and political hegemony has been dictating life in the Arab city for decades, leading to inequality and suffering, with youth being hit the hardest. Policies that prioritize profit-making, and discriminate against people from lower social classes and disempowered ethnicities, can serve to catalyze these groups to assert their power and find their space within their cities. This means that cities, as much as they are sites of struggles and inequality, can be centers of revolution. Urban activism can take several forms, among them, trade unionism, neighborhood initiatives, grassroot and civil-based organizations advocacy, mass mobilizations and protests. The co-opting of trade union leaders has rendered unions a bureaucratic arm of the current regimes, instead of being channels that help workers fight for their rights. Civil rights organizations work is limited, and they are also busy finding donor funding and, many times, end up working on projects dictated by funders, rather than the actual needs of the people. In Arab cities Bayat (2002) referred to a pervasive form of urban activism, which he referred to as the "quiet encroachment of the ordinary," which he defines as the silent pervasive advancement on spaces in the city, through unlawfully acquiring land, building homes and getting urban services, jobs or business opportunities in a quiet and prolonged fashion. The idea is that if current structures and bureaucratic systems that rule the city are exclusionary, then people will start creating their own law and their own reality. State and municipal authorities fight this form of resistance fiercely.

Youth are not passive recipients of state policies, they have found their own channels to express their outrage at the state of their marginalization. Although they are both pushed out and self-excluded from current formal political structures in their countries, they are still active informally on social media. Their online activism has also been translated into mass street protests, where they have claimed their space at the centers of their cities. Arab youth have reclaimed those central urban spaces during the uprisings, not only by organizing protests, but also through visual art. The walls of Arab cities were painted with messages of defiance. Today, over a decade after the uprisings, the social and economic ailments that pushed the youth to protest in their city centers are still present, which means unrest is likely to erupt again.

Bibliography

Abdelrahman, M. (2013). Ordering the disorderly? Street vendors and the developmentalist state. *Jadaliyya*. www.jadaliyya.com/pages/index/9542/ordering-the-disorderly-street-vendors-and-the-dev

Abdelrahman, M. (2017). Policing neoliberalism in Egypt: The continuing rise of the 'securocratic' state. *Third World Quarterly, 38*(1), 185–202.

Abdselm, H. (2021). *The "wrong generation" leads Tunisia's protests*. Available at https://carnegieendowment.org/sada/84596. Accessed 25 March 2021.

Ahram. (2020). *Egypt demolishes thousands of illegal buildings in 3 months—politics—egypt*. https://english.ahram.org.eg/NewsContent/1/64/374066/Egypt/Politics-/Egypt-demolishes-thousands-of-illegal-buildings-in.aspx. Accessed 20 March 2022.

Al Haq. (2021). In repression of Palestinian Assembly, Israel instigates a campaign of collective punishment in Jerusalem. Available at: https://www.alhaq.org/advocacy/18478.html. Accessed 29 October 2022.

Allegra, M., Bono, I., Rokem, J., Casaglia, A., Marzorati, R., & Yacobi, H. (2013). Rethinking cities in contentious times: The mobilisation of urban dissent in the 'Arab Spring.' *Urban Studies, 50*(9), 1675–1688.

Al Jazeera. (2021). *Palestinian solidarity protests held around the world.* https://www.aljazeera.com/news/2021/5/22/palestinian-solidarityprotests-marked-around-the-world. Accessed 20 May 2022.

Al Shalabi, J. (2011). Jordan: Revolutionaries without a revolution. *Confluences Méditerranée, 77*(2), 91.

Aman, A. (2016). Cairo tries to reassure citizens and Israel amid island controversy. *Al-Monitor: The Pulse of the Middle East.* [online]. Available at https://www.almonitor.com/originals/2016/04/egypt-saudi-arabia-islands-controversy-salman-israel.html. Accessed 15 March 2021.

Amnesty International. (2022). Palestine: Authorities have failed to ensure accountability for the killing of Nizar Banat. Available at https://www.amnesty.org/en/latest/news/2022/06/palestineauthorities-have-failed-to-ensure-accountability-for-the-killing-of-nizar-banat/. Accessed 29 June 2022.

Amin, S. (2020). *Artistic freedom of expression shrinks in 'new' Egypt.* Al-Monitor: The Pulse of the Middle East. https://www.almonitor.com/originals/2020/01/january-25-anniversary-freedom-of-expression.html#ixzz7LnIROSNM Accessed 15 March 2021.

Andrews, F. (2020). *Banksy in Palestine: A look at the street artist's work in Gaza and the West Bank.* Available at https://www.thenationalnews.com/arts-culture/art/banksy-in-palestine-a-look-at-the-street-artist-s-work-in-gaza-and-the-west-bank-1.1031618. Accessed 25 March 2021.

Barthel, P. (2010). Arab mega-projects: Between the Dubai effect, global crisis, social mobilization and a sustainable shift. *Built Environment, 36*(2), 133–145.

Bayat, A. (2002). Activism and social development in the Middle East. *International Journal of Middle East Studies, 34*(1), 1–28.

Bayat, A. (2015). Plebeians of the Arab Spring. *Current Anthropology, 56*(S11), 33–43.

Beaumont, P. (2017). Israeli security forces and Palestinian worshippers clash outside Al-Aqsa Mosque. *The Guardian. Guardian News and Media.* Available at https://www.theguardian.com/world/2017/jul/27/israel-removes-further-securitymeasures-from-al-aqsa-compound. Accessed 29 October 2022.

Bellin, E. (2012). Reconsidering the robustness of authoritarianism in the Middle East: Lessons from the Arab Spring. *Comparative Politics, 44*(2), 127–149.

Ben Yahmed, Z., & Yerkes, S. (2018). Tunisians' revolutionary goals remain unfulfilled. Available at https://carnegieendowment.org/2018/12/06/tunisians-revolutionary-goals-remain-unfulfilled-pub-77894. Accessed 15 March. 2021.

Betselem. (2020). *Palestinians killed by Israeli security forces in the West Bank, since Operation Cast Lead.* https://www.btselem.org/statistics/fatalities/after-cast-lead/by-date-of-death/westbank/palestinians-killed-by-israeli-security-forces. Accessed 15 March 2021.

Betselem. (2021). *East Jerusalem.* Betselem: The Israeli Information Center for Human Rights in the Occupied Territories. https://www.btselem.org/topic/jerusalem. Accessed 15 May 2021.

Blackburn, D. (2018). Focus: Trade unions and democracy in Egypt. *International Union Rights, 25*(2), 10.

Bogaert, K. (2013). Contextualizing the Arab revolts: The politics behind three decades of neoliberalism in the Arab world. *Middle East Critique, 22*(3), 213–234.

Bush, R. (2004). Poverty and neo-liberal bias in the Middle East and North Africa. *Development and Change, 35*(4), 673–695.

Calder, M., MacDonald, R., Mikhael, D., Murphy, E., & Phoenix, J. (2017). Marginalization, young people in the South and East Mediterranean, and policy: An analysis of young people's experiences of marginalization across six SEM countries, and guidelines for policy-makers (Power2Youth Working Paper No. 35).

Cantini, D. (2012). Discourses of reforms and questions of citizenship: The university in Jordan. *Revue Des Mondes Musulmans Et De La Méditerranée, 131*, 147–162.

Carnegie Endowment for International Peace. (2010). *Kefaya.* https://carnegieendowment.org/2010/09/22/kifaya-pub-54922. Accessed 15 May 2021.

Connell, T. (2013). *Solidarity Center—Jordanian unions establish independent federation.* Solidarity Center. https://www.solidaritycenter.org/jordanian-unions-establish-independent-federation/

Connell, T. (2019). *Solidarity Center—Global Unions Urge Jordan to Withdraw Harsh Labor Laws.* Solidarity Center. https://www.solidaritycenter.org/global-unions-urge-jordan-to-withdraw-harsh-labor-laws/. Accessed 1 June 2021.

De Soto, H. (2011). The free market secret of the Arab revolutions. *Financial Times.* https://www.ft.com/stream/890e44f4-5fe3-3184-ab06-cbc3d27d0b85?page=41. Accessed 1 May 2021.

Domaradzka, A. (2018). Urban social movements and the right to the city: An introduction to the special issue on urban mobilization. *VOLUNTAS: International Journal of Voluntary and Nonprofit Organizations, 29*(4), 607–620.

Dreisbach, T., & Smadhi, A. (2015). *Illegal street vendors struggle for survival in post-revolution Tunisia*. Middle East Eye. Available at https://www.middleeasteye.net/fr/news/illegal-street-vendors-struggle-survival-post-revolution-tunisia-1281841878. Accessed 1 May 2022.

Dwonch, A. (2019). *Palestinian youth activism in the internet age: Online and offline social networks after the Arab Spring*. SOAS Palestine Studies. I.B. Tauris.

El Sharkawy, S., & El Agati, M. (2021). *Independent trade unions: Between political developments and internal factors—Egyptian case study 2004–2015*. Retrieved May 27, 2022, from https://www.arab-reform.net/publication/independent-tradeunions-between-political-developments-and-internal-factors-egyptian-case-study-2004-2015/

Espinosa, A. (2018). *Translocal urban activists: Brokers and the geographies of urban social movements* (Urban Studies Master Thesis). Malmo University.

ETUF. (2022). *The establishment of the union*. https://www.etufnews.com/-نشأةالاتحاد-العامعام . Accessed 20 March 2022.

Ginwright, S., Cammarota, J., & Noguera, P. (2005). Youth, social justice, and communities: Toward a theory of urban youth policy. *Social Justice, 32*(3 (101)), 24–40.

Handala. (n.d.). *Through the eyes of a Palestinian refugee*. Available at http://www.handala.org/index.html. Accessed 15 May 2022.

Harris, M. (2015). *Jordan's youth after the Arab Spring*. https://www.jstor.org/stable/pdf/resrep10167.pdf?refreqid=excelsior%3A930410c20dee9994babaee809721cb01&ab_segments=&origin=. Accessed 1 May 2021.

Hibou, B. (2011). *The force of obedience*. Polity Press.

ICAHD. (2022). The Israeli committee against house demolitions. *Homepage*. https://icahd.org/. Accessed 5 March 2022.

ITUC. (2019). *Trade union development projects directory*. https://projects.ituc-csi.org/gfjtu. Accessed March 27, 2022.

Jordan Labor Watch. (2012). *Freedom of association in Jordan: The new trade unions, controversy of legitimacy*. https://library.fes.de/pdf-files/bueros/amman/10103.pdf. Accessed 1 May 2022.

Lau, L. (2013). *The murals of Mohammad Mahmoud Street: Reclaiming narratives of living history for the Egyptian people | writing program*. https://www.bu.edu/writingprogram/journal/past-issues/issue-5/lau/. Accessed 1 May 2021.

Lim, M. (2012). Clicks, cabs, and coffee houses: Social media and oppositional movements in Egypt, 2004–2011. *The Journal of Communication, 62*(2), 231–248.

Mandour, M. (2021). *The sinister side of Sisi's urban development*. Carnegie Endowment for International Peace. https://carnegieendowment.org/sada/84504. Accessed 15 February 2022.

Mansoor, S. (2021). Muna and Mohammed El-Kurd are on the 2021 time 100 list. *Time*. Available at https://time.com/collection/100-most-influential-people-2021/6096098/muna-mohammed-el-kurd/. Accessed 29 September 2022.

Miller, B., & Nicholls, W. (2013). Social movements in urban society: The city as a space of politicization. *Urban Geography, 34*(4), 452–473.

Nagati, O., & Stryker, B. (2013). *Archiving the city in flux. Cairo's shifting urban landscape since the January 25th revolution*. https://issuu.com/clustercairo/docs/archiving_the_city_in_flux. Accessed 1 May 2021.

Netterstrøm, K. (2016). The Tunisian general labor union and the advent of democracy. *The Middle East Journal, 70*(3), 383–398.

Pelham, N. (2011). *Jordan's Balancing Act—MERIP*. Available at https://merip.org/2011/02/jordans-balancing-act/. Accessed 16 May 2022.

Peters, A., & Moore, P. (2009). Beyond boom and bust: External rents, Durable authoritarianism, and institutional adaptation in the Hashemite Kingdom of Jordan. *Studies in Comparative International Development, 44*(3), 256–285.

Rabie, H. (2019). *Egypt plans to reinstate municipal oversight councils*. Al-Monitor: The Pulse of the Middle East. Retrieved 9 May 2022, from https://www.al-monitor.com/originals/2019/10/egypt-local-council-elections-municipalities-corruption.html

Reporters Without Borders. (2021) *Less press freedom than ever in Egypt, 10 years after revolution*.

Roth, K. (2021, January 13). *Rights trends in Jordan*. Human Rights Watch. Retrieved October 20, 2022, from https://www.hrw.org/world-report/2021/country-chapters/jordan.

RSF. (2021). https://rsf.org/en/less-press-freedom-ever-egypt-10-years-after-revolution. Accessed 1 May 2021.

Schwedler, J. (2018). *Jordan's austerity protests in context. Atlantic Council*. Available at https://www.atlanticcouncil.org/blogs/menasource/jordan-s-austerity-protests-in-context/. Accessed 29 May 2022.

Sims, D. (2014). *Egypt's desert dreams: Development or disaster?* (1st ed.). The American University in Cairo Press.

Susser, A. (2021). *Still standing, but shaky: Jordan at 100*. https://fathomjournal.org/still-standing-but-shakyjordan-at-100/. Accessed 26 December 2021.

Tobin, S. (2012). Jordan's Arab spring: The middle class and anti-revolution. *Middle East Policy, 19*, 96–109.

Tripp, C. (2013). *The power and the people: Paths of resistance in the middle east*. Cambridge University Press.

Uitermark, J., Nicholls, W., & Loopmans, M. (2012). Cities and social movements: Theorizing beyond the right to the city. *Environment and Planning a: Economy and Space, 44*(11), 2546–2554.

Vohra, A. (2022). Why Israel is afraid of Palestinian funerals. *Foreign Policy*. Available at https://foreignpolicy.com/2022/05/25/shireen-abu-akleh-funeral-israeli-palestinian-conflict-journalist-killing/. Accessed 29 October 2022.

Wated, M. (2022). One year on, Haifa's uprising is inspiring a united Palestinian movement. *Middle East Eye*. Available at https://www.middleeasteye.net/news/palestine-israel-haifa-karamah-uprising-oneyear. Accessed 29 May 2022.

World Report. (2021). Rights trends in Jordan. (2021). https://www.hrw.org/world-report/2021/country-chapters/jordan. Accessed 25 March 2022.

Yee, V. (2022). 'A slow death': Egypt's political prisoners recount horrific conditions. *The New York Times*. Available at https://www.nytimes.com/2022/08/08/world/middleeast/egypts-prisons-conditions.html. Accessed 29 October 2022.

Yerkes, S. (2017). *Where have all the revolutionaries gone?* Brookings. https://www.brookings.edu/wp-content/uploads/2017/03/cmep_20160317_where_have_revolutionaries_gone.pdf. Accessed 15 May 2022.

Zemni, S. (2017). The Tunisian revolution: Neoliberalism, urban contentious politics and the right to the city. *International Journal of Urban and Regional Research, 41*(1), 70–83.

CHAPTER 7

Conclusion

For the last several decades, the logic that has shaped Arab cities is capitalism and political hegemony, leading to the marginalization of people from lower social classes and disempowered ethnicities. The cities entered into a competitive logic to maximize economic growth and capital accumulation, which prioritized tourist projects, foreign investments in real estate projects, luxury resorts, and industrial zones (Bogaert, 2013), at the expense of the majority of city inhabitants. These projects did not result in expanding private sector jobs, as claimed by their advocates, and only resulted in the hiking up of prices on middle-class and poor segments of society, pushing them further towards the periphery, or sometimes, integrating them within these projects as cheap labour, with precarious work conditions, keeping them below the poverty line. Arab youth, in particular, are paying the price of these policies, with higher levels of unemployment and precarious working and living conditions. Youth are faced with a different reality than their parents. Although this youth cohort is more educated, have higher expectations, and are more exposed to the world, what is available to them in the market, in terms of quantity or quality of job opportunities is limited (Calder et al.). They find themselves jumping from one temporary job to another, they are forced to spend a lot of time applying for jobs and filling forms, to find themselves in jobs that do not meet their credentials, with low salaries and

R. A. Nuseibeh, *Urban Youth Unemployment, Marginalization and Politics in MENA*, Middle East Today,
https://doi.org/10.1007/978-3-031-15301-3_7

no benefit packages (Standing, 2014). Instead of the labour market being another venue for youth to build their capabilities and advance their skills, it is a place where their labour is exploited for minimal returns, on both the financial and the skills development sides. The education sector is also failing youth. In our four contexts (Jerusalem, Tunis, Cairo and Amman), there have been great achievements in terms of enrollment in education. However, there are also several problems in the education sector with regards to pre-primary education enrollment, and the quality of education provided. In addition, there are issues of disparity of opportunity, along social class and ethnic lines, as well as problems with infrastructure, and teachers' qualifications. The authoritarianism and patriarchy by which the states are governed has seeped into the classroom environment, with limited educational programs that arouse students' curiosities, understanding of opposing views, and freedom to critically analyze or to doubt the current structures. The governance of the education systems, in the four contexts, does not reflect a democratic approach that involves grassroot organizations, students' committees, parents' committees and local educators and academics. Education policy is mostly a top-down approach with semi-decentralization attempts. Even when there are policies and legislations that support decentralization, local authorities and schools lack the capacities and the resources to actually implement them in a meaningful way. Neoliberal policies have also affected institutions of higher education; this can be witnessed in the push for universities to assume roles as economic drivers.

Exclusion of youth in cities, in the education sector, in the economy, and from any meaningful political participation , has catalyzed these groups to assert their power and find their space within their cities. Cities have become sites of both inequality and marginalization and spaces of revolution and dissent. Urban activism has taken several forms, such as online activism and lobbying, to mobilization of protesters, and the quiet reclaiming of spaces in the city outside the laws imposed by municipal and state authorities. Since the current structure and bureaucratic systems that rule the city are exclusionary, people are starting to create their own law and their own reality. Youth are not passive recipients of state policies, they have found their own channels to express their outrage at the state of their marginalization. Although they are both pushed out and self-excluded from formal political structures in their countries, they are still active informally on social media.

This book has called for re-imagining the city through the "right to the city" and the "capabilities approach" frameworks. Imagining the cities as not only fulfilling the social, economic, civil, and political rights of all residents, but also empowering the city residents to participate in the shaping of their cities. Addressing injustice in the city through the "capabilities approach" can be a starting point to assess how certain groups are disadvantaged and what can be done to empower them to be part of the decision-making institutions. As Deneulin (2014) explains, first we should identify which valuable capabilities people are deprived of, and then discuss the appropriate remedies for the injustice, and the exclusion they face. There is no perfect scenario for each city; each city has its own issues that need to be addressed specifically, and the urban dwellers are the only ones capable of finding those solutions. The "capabilities approach" provides the tools for wellbeing evaluation within the framework of the "right to the city"; it shows how different dimensions of wellbeing can affect each other (Deneulin, 2014). For example, sexual harassment and lack of safety can affect women's movement within the city; the lack of a safe work environment can also affect their employment prospects. Another contribution of the "capability approach" to the "right to the city" is the potential for evaluating the role of insititutions in facilitating or constraining the realization of the capabilities of urban residents (Deneulin, 2014). This calls us to review these institutions and their roles in society; asking questions which include: are political institutions democratically elected? are they channels that implement the needs of the city residents equally? how do the urban residents relate to the urban municipality etc. The same goes for institutions of education, their governance, their outreach, and the services they provide (ibid.) We need to ask wether these institutions are building capabilities or depriving capabilities (Unterhalter, 2003). This book has demonstrated that neither the political nor the education institutions in the city are serving people equally.

Therefore, we need to re-imagine the governance of the city. The autonomy of local governments in cities makes them more able to implement creative projects that cater to the local issues and needs of their communities, examples from around the world show that democratically elected local governments are more efficient in serving the needs of the city residents. Authoritarian installation of local officials, top-down policy or a replica of other cities' policies, rather than local policies, that give space for grassroot participation and residents participation, do not result in cities serving all their residents/citizens. As long as people in the four cities we studied are not participating in the decision-making processes of how their cities are shaped, governed and how resources are allocated, inequality and segregation will persist.

Re-imagining the labour markets, to create a more inclusive environment that can provide space for the least advantaged groups, such as women, youth, and people residing in disadvantaged neighborhoods. Creating government and private sector synergies that are functioning in a democratic environment, where policies are transparent, and can be subject to the people's scrutiny. As long as the media is stifled and people's freedom to speak and protest are restricted, governments can get away with corrupt authoritarian policies, that benefit the few; political elite will keep benefiting from their connections to secure monopolies, and economic reforms will remain superficial. In democratic societies "the right to the city" is demanded through social practices embedded within democratic expression (Al-Hamarneh, 2019). This can be seen in grass-root movements and activists demanding deeper participation in decision making processes that shape their cities, and defending their urban rights, through the use of democratic tools, such as freedom of expression, freedom of assembly, protesting, lobbying, holding local officials accountable etc. (ibid.). In Arab cities these forms of expression are suffocated, and when there are expressions of urban demands that do not conform to the regime's ideology and policies, they are criminalized and crushed.

Reimagining the role of education to make it more democratically governed, and more inclusive in implementation. Starting from pre-primary education, which is not mandatory, and is very limited in scope, and most provisions are tuition fee-based. Although it has been established that investing in quality early preschool education is extremely important, not just for its benefits on the individual and his/her educational trajectory, but also for the social and the economic health of the communities as a whole (Pianta et al., 2009), still provisions are limited in our four contexts, and budgets are mostly directed to institutions of higher education, where mostly upper-class students are represented. There is also a pressing need to overhaul primary and secondary school education systems by improving infrastructure, teachers training, advancing the use of technology and upgrading curriculums to be more inclusive and gender-sensitive. However, all these changes cannot be done in the current authoritarian governance of our four contexts. More autonomy in governance is needed on the local level, where educators take ownership of their schools and regions, and get rid of authoritarian state bureaucracies. However, autonomy without support will be a failure; local actors need to be empowered and supported with capacity building and budgets, in order for this to yield successful results. This applies to the governance of institutions of higher education, which cannot achieve academic excellence, and be forces of change in society, pushing for inclusion and equality, as long as their leadership is an extension of authoritarian regimes. There is a need to re-imagine the governance of educational institutions on various levels, in order to achieve more democratic structures of teaching and learning. In their current state, these institutions that are meant to be spaces for youth to learn, grow, express their grievances, debate and change realities, are failing them.

In the current climate, and as we have seen from the interviews with youth, they are frustrated with the opportunities available to them in their cities, whether educational or in the labour market. They are also not empowered to be part of the decision-making apparatuses, and to have the power to change. Addressing the issue of youth exclusion in cities requires an in-depth look into their civil, political and social rights in the city, along intersectional lines of gender, ethnicity, and social class. This will enable us to understand where certain groups suffer from capability deprivation and then work on remedying this injustice. However, without a true democracy and widespread corruption, intervention programs will remain superficial and youth will remain a marginalized group.

Bibliography

Al-Hamarneh, A. (2019). Right to the city in the Arab World: Case studies from Amman and Tunis metropolitan areas. *Neoliberale Urbanisierung: Stadtentwicklungsprozesse in der arabischen Welt*, 185–214.

Bogaert, K. (2013). Contextualizing the Arab revolts: The politics behind three decades of neoliberalism in the Arab world. *Middle East Critique, 22*(3), 213–234.

Deneulin, S. (2014). *Creating more just cities: The right to the city and capability approach combined* (Bath Papers in International Development and Wellbeing, No. 32). University of Bath, Centre for Development Studies (CDS).

Pianta, R., Barnett, W., Burchinal, M., & Thornburg, K. (2009). The effects of preschool education. *Psychological Science in the Public Interest, 10*(2), 49–88.

Standing, G. (2014). Understanding the Precariat through labour and work. *Development and Change, 45*(5), 963–980.

Unterhalter, E. (2003). Education, Capabilities and Social Justice. *UNESCO*. Available at https://unesdoc.unesco.org/ark:/48223/pf0000146971. Accessed 4 October 2020.

Index

R. A. Nuseibeh, *Urban Youth Unemployment, Marginalization and Politics in MENA*, Middle East Today,
https://doi.org/10.1007/978-3-031-15301-3